FESTKÖRPERPROBLEME

ADVANCES IN SOLID STATE PHYSICS 33

FESTKÖRPER PROBLEME

ADVANCES IN SOLID STATE PHYSICS 33

Edited by
Reinhard Helbig

Editor:

Prof. Dr. Reinhard Helbig
Institut für Angewandte Physik
Universität Erlangen - Nürnberg
Staudtstr. 7
D-91058 Erlangen

Originally published by Friedr. Vieweg & Sohn Verlagsgesellschaft mbH, Braunschweig/Wiesbaden in 1994.

Cover design: Barbara Seebohm, Braunschweig
Printed on acid-free paper

ISBN 978-3-662-16073-2 ISBN 978-3-540-75339-1 (eBook)
DOI 10.1007/978-3-540-75339-1

ISSN 0430-3393

Foreword

Rich in tradition the series "Advances in Solid State Physics" published a collection of review articles about actual results in solid state physics every year. For many young an old scientists these reviews were a first introduction into a new field or useful to get a general overview. Normally the published articles were selected from plenary and invited talks of the spring meeting on solid state physics of the German Physical Society. As an exception in 1993 the spring meeting was organized together with the European Physical Society and the proceedings will be published somewhere else. Therefore we have invited some colleagues to publish a review article (of course together with own results) in the present volume 33 of the "Advances in Solid State Physics".

Erlangen, December 1993 R. Helbig

Contents

Reconstruction of the Cleavage Faces of Tetrahedrally Coordinated Compound Semiconductors

C. B. Duke

Xerox Webster Research counter, 800 Phillips Road, 0114-38D, Webster, New York 14580 USA

Summary: Tetrahedrally coordinated compound semiconductors occur in two crystallographic allotropes: zincblende and wurtzite. Zincblende materials exhibit a single cleavage face: The (110) surface consisting of equal numbers of anions and cations which form zig-zag chains directed along <110> directions in the surface. Wurtzite materials exhibit two cleavage faces, both consisting of equal numbers of anion and cation species. The $(10\bar{1}0)$ cleavage surfaces consist of isolated anion-cation dimers back bonded to the layer beneath whereas the $(11\bar{2}0)$ surfaces consist of anion-cation chains, analogous to those on zincblende (110) but with four rather than two inequivalent atoms per surface unit cell. All three surfaces exhibit reconstructions which do not alter the symmetry of the surface unit cell but which lead to large ($\approx$ 1 Å) deviations of the positions of the atomic species in the uppermost layer(s) from those in the truncated bulk solid. These reconstructed surface geometries have been determined quantitatively for the (110) surfaces of zincblende structure AlP, AlAs, GaP, GaAs, GaSb, InP, InAs, InSb, ZnS, ZnSe, ZnTe and CdTe; the $(10\bar{1}0)$ surfaces of wurtzite structure ZnO and CdSe; and the $(11\bar{2}0)$ surfaces of CdSe. Theoretical predictions of these reconstructed geometries have been given which are in either quantitative or semiquantitative correspondence with the experimentally determined structures. Analysis of the trends exhibited by the members of each class of cleavage surface and comparison thereof with theoretical predictions permit the extraction from these results of generalizations characteristic of novel types of surface chemical bonding. The most important of these is the notion that for each class of surface the atomic geometries are approximately "universal" when their coordinates are properly scaled with the bulk lattice constant. A quantitative description of this result is presented which reveals that extensions of the concepts of inorganic molecular coordination chemistry are required to predict the cleavage-surface atomic geometries and electronic structures of binary tetrahedrally coordinated compound semiconductors.

1 Introduction

A study of the reconstructed atomic geometries of the cleavage faces of tetrahedrally coordinated compound semiconductors is of particular interest for three reasons. First, for semiconductors crystallizing in the zincblende structure, a rather complete account of the structures of the (110) cleavage surfaces is available for sp^3-bonded binary compounds, i.e., AlP, AlAs, GaP, GaAs, GaSb, InP, InAs, InSb, ZnS, ZnSe, ZnTe, and CdTe [1-3]. Therefore the systematics of the variations of these structures from one material to another can be determined and interpreted. Second, after considerable controversy over a period of nearly a decade, the atomic geometry of GaAs(110), the benchmark of compound semiconductor surface structures, seems to have been determined definitively utilizing a wide variety of experimental techniques including low-energy electron diffraction (LEED) [4-6], ion scattering spectroscopy [7], ion channelling spectroscopy [8], He atom diffraction [9], scanning tunnelling microscopy [10], and secondary ion mass spectroscopy [11]. Studies of the filled electronic surface states by photoemission [12] and of empty surface states by inverse photoemission [13,14] have been utilized in attempts to distinguish between structural models, the most recent results obtained using both methods [12,14] being compatible with the accepted $\omega_1 = 29°$ bond-rotated model of the atomic geometry of GaAs(110). Third and finally, the study of the systematics of the atomic geometries of the zincblende (110) cleavage surface as well as the wurtzite $(10\bar{1}0)$ and $(11\bar{2}0)$ cleavage faces has revealed new and unexpected phenomena. For example, in spite of very different small-molecule coordination chemistries the III-V and II-VI compounds were found to exhibit approximately "universal" surface structures when the structural parameters are scaled linearly with the bulk lattice constant for both zincblende [15] and wurtzite [3,16] materials. This remarkable fact, which stands in contradiction both with models based upon local coordination chemistry (17) and with expectations that ionicity might govern trends in the surface structures [18,19], can be understood only in terms of a new type of topologically dominated surface rehybridization in which the effects of the surface template dominate those of the local coordination chemistry [3,20,21].

Studies of the surfaces of tetrahedrally coordinated compound semiconductors have a long and venerable history. Initially motivated by early work at Bell Laboratories [22], extensive LEED and work function measurements were reported during the period 1964-75. These measurements, reviewed in 1975 by Mark et al. [23], revealed that the non-polar cleavage surfaces were more stable than the low-index polar surfaces insofar as their symmetry parallel to the surface is identical to that characteristic of the bulk solid whereas in the case of polar surfaces the symmetry is lowered (i.e., the surface unit cell is larger than that in the bulk). Non-polar surfaces contain equal numbers of anion and cation species whereas polar surfaces contain an excess of one type of species as illustrated in Figs.1-3 for the non-polar (110) and polar (100) and (111) surfaces of zincblende-structure binary semiconductors. The reduction in the surface symmetry of the polar surfaces was ascertained from the symmetries of

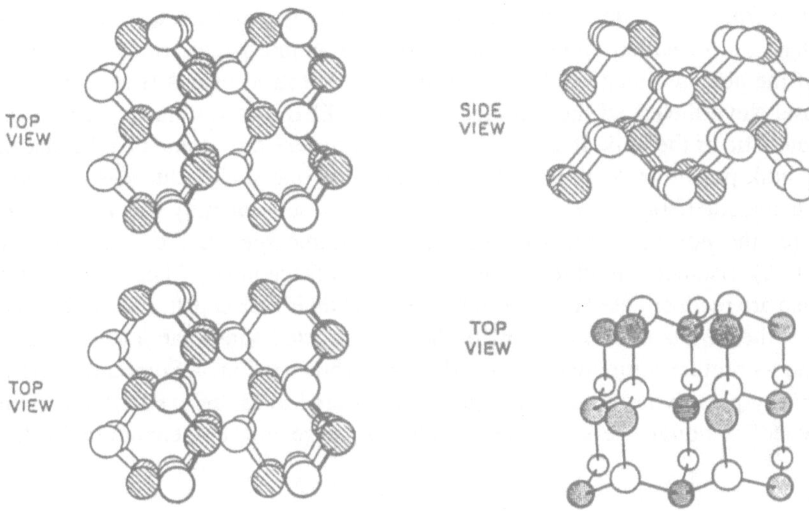

Figure 1
Schematic indication of the truncated bulk geometry of the non-polar (110) cleavage faces of zincblende structure compound semiconductors. Shaded circles indicate anions whereas open circles indicate cations. (Adapted from Duke [168].)

Figure 2
Schematic indication of the truncated bulk geometry of the polar (100) surfaces of zincblende structure compound semiconductors. Shaded circles indicate anions whereas open circles indicate cations. (Adapted from Duke [25].)

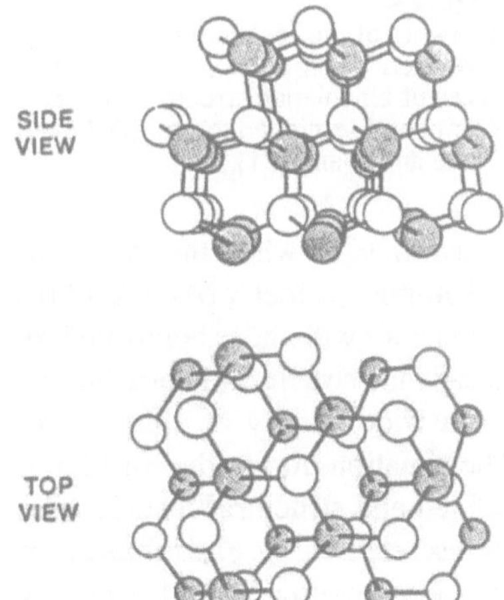

Figure 3
Schematic indication of the truncated bulk geometry of the polar (111) surfaces of zincblende structure compound semiconductors. Shaded circles indicate anions whereas open circles indicate cations. (Adapted from Duke [25].)

the LEED beams, typically displayed on a fluorescent screen as a "LEED pattern" using post-diffraction acceleration electron optics [22,23]. Such an apparatus gives no quantitative information about the actual surface atomic geometries because an analysis of the absolute magnitudes of the diffracted electron beams is required for that purpose. Nevertheless, if the symmetries of the LEED patterns are reduced from that characteristic of the bulk, it can inferred that the surface atoms must be displaced from their bulk positions, leading to the nomenclature that such reduced-symmetry patterns are characteristic of "reconstructed" surfaces. It subsequently was found [24] that even for the non-polar surfaces, the surface atomic species are displaced by large (≈ 1 Å) distances relative to the truncated bulk geometry. Therefore these surfaces also are reconstructed in spite of the fact that their LEED patterns exhibit the symmetry of the bulk lattice. A drawing of the reconstructed non-polar (110) surfaces of zincblende-structure compound semiconductors is shown in Fig. 4. Sometimes these bulk-symmetry-preserving surface structures are referred to as "relaxed" rather than "reconstructed" although we shall use the latter more generic nomenclature in this article.

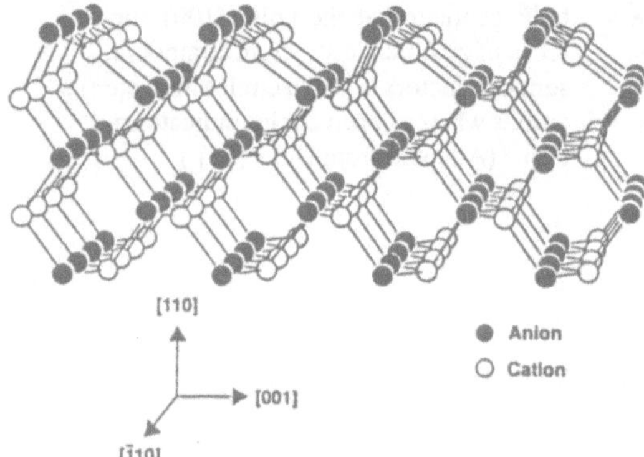

Figure 4
Drawing of the reconstructed ("relaxed") non-polar (110) cleavage faces of zincblende structure binary compound semiconductors. (Adapted Duke and Wang [21].)

A new era in semiconductor surface science dawned in 1976 when the theory of LEED had advanced sufficiently that the actual surface atomic geometry of GaAs(110) could be determined by comparing measured LEED intensities with those computed for various structural models [24]. During the ensuing decade the structure-determination methodology was refined, improved, and applied to a wide variety of zincblende-structure binary III-IV and II-VI semiconductors. The situation during the middle of this period, when the techniques were still being refined and structural results were fragmentary (and sometimes controversial), may be ascertained by examination of reviews by Duke [25] and by Mark et al. [26] which were written at that time. By 1983 a significant body of quantitative surface structures for different semiconductors had been accumulated for comparison with model predictions [2,15]. Controversies remained, which were resolved over time [4-6]. By 1988 an extensive collection of structural determinations for the (110) cleavage surfaces of binary III-V and II-VI semiconductors had been established [1].

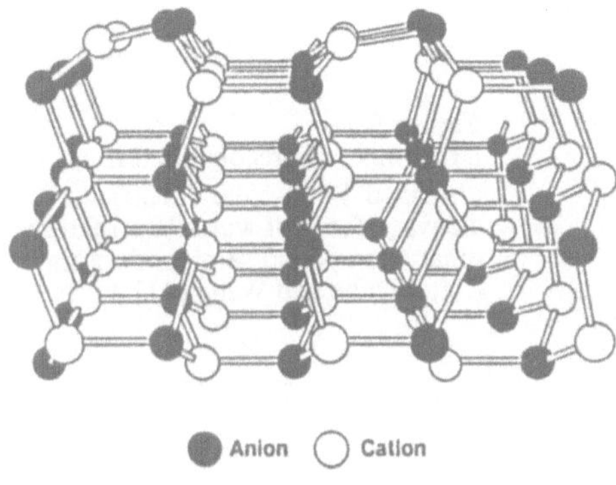

Figure 5
Drawing of the reconstructed ("relaxed") nonpolar $(10\bar{1}0)$ cleavage faces of wurtzile structure binary compound semiconductors. (After Due and Wang [21.])

As noted earlier, the body of "experimental" structures for zincblende (110) surfaces proved inconsistent with theoretical expectations based on both local coordination chemistry [17] and ionicity-structure correlations [18], thereby stimulating a re-examination of the theory of chemical bonding at the surfaces of sp^3 bonded compound semiconductors, especially II-VI materials [3,27]. This re-examination led to the identification of a pseudo-Jahn-Teller surface-state lowering mechanism, associated with an activationless bond-length-conserving rotation of the surface species, as the driving force of a "universal" reconstruction of the (110) cleavage surfaces of both III-V and II-VI sp^3-bonded zincblende structure compound semiconductors [3,21,28]. Since the same concepts should be applicable to wurtzite-structure materials, a major effort was undertaken to extend both the model calculations [16, 21] and experimental structure determinations [29-32] to these systems. Analogous "universal" reconstructions were predicted for both the $(10\bar{1}0)$ (Fig. 5) and $(11\bar{2}0)$ (Fig 6) surfaces of wurtzite-structure II-VI compounds. Experimental evidence is still fragmentary, but the results available to date on the $(10\bar{1}0)$ surfaces of ZnO (29] and CdSe [30-32] as well as CdSe $(11\bar{2}0)$ [31,32] confirm the expectations based on these predictions, suggesting that more surprises might be uncovered in studies of II-VI surface chemistry.

Our purposes in this review are to survey the scope of the quantitative experimental determinations of the atomic geometries of the cleavage faces of compound semiconductors, the extent to which the resulting structures have been shown to be compatible with theoretical predictions and with measured electronic and atomic excitation spectra, and the new concepts concerning surface chemical bonding which have emanated from these efforts. We proceed by first examining the extensively-studied zincblende (110) cleavage surfaces and then considering the available results for the wurtzite $(10\bar{1}0)$ and $(11\bar{2}0)$ surfaces.

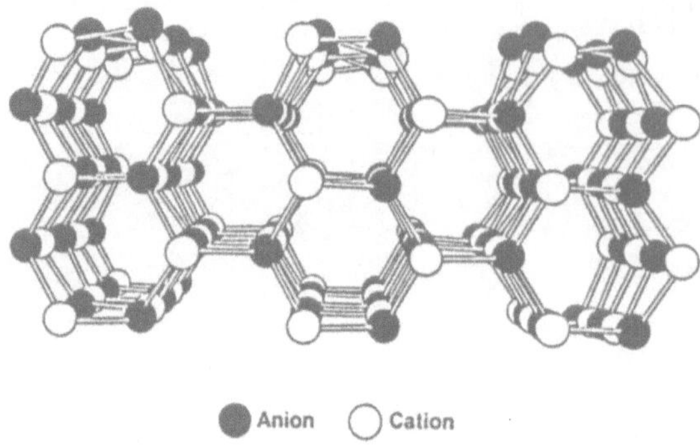

Figure 6
Drawing of the reconstructed ("relaxed") non-polar (11$\bar{2}$0) cleavage faces of wurtzile structure binary compound semiconductors. (After Duke and Wang [21].)

2 Zincblende (110)

Studies of the atomic geometries of the (110) surfaces of zincblende-structure compound semiconductors are the benchmarks in semiconductor surface structure determination. A wide variety of experimental techniques have been applied for this purpose, with early LEED results being confirmed repeatedly [1,2,6,27]. Since the field has been reviewed extensively, we proceed by extracting the highlights from the most recent published reviews [1,2,27] and augmenting this discussion with a more extensive treatment of the major new results since the pre 1986 period covered by these reviews. From an experimental perspective, these results encompass new confirmations of previously proposed geometries of the (110) surfaces of GaAs [33], InSb [34,35] and CdTe [36-38]. In addition, surface vibrational modes have been observed [39,40] and interpreted [28,41] for GaAs (110). From a theoretical perspective, three advances have occurred since 1985. First, tight-binding total energy methods have been extended to II-VI compounds and applied to predict the atomic geometries and electronic structures of the (110) surfaces of ZnS [42], ZnSe [43] and CdTe [44]. Second, more accurate *ab initio* pseudopotential methods have been applied to calculate the surface atomic geometries of GaAs(110) [45, 46] and ZnSe(110) [47]. Third, a series of recent theoretical calculations [19, 48, 49] has been invoked to resurrect the old issue [23, 25] of the dependence of surface structure on ionicity. Such a dependence had been proposed in the mid 1970's [18,50,51] and shown to be incompatible with a variety of zincblende (110) surface structure determinations [15,52,53]. Finally, it recently was discovered [54] that bismuth forms epitaxical overlayers on GaAs(110), and a number of studies of this system have been reported [54-58].

2.1 Atomic Geometry – Experimental Determination

The most recent comprehensive review of the experimental determinations of the atomic geometries of the (110) surfaces of zincblende-structure compound semiconductors is that given by Duke [1]. The reconstructed surfaces exhibit the same unit cell dimensions as the truncated bulk lattice, but the species in the upper-most two (and probably deeper) atomic layers are displaced for their bulk positions. When a semiconductor surface is formed, bonds are broken. This fact both puts the surface under stress and relaxes the bulk constraints on atomic movements to relieve this stress. A universal response to this situation by tetrahedrally coordinated compound semiconductors is to form new types of chemical bonds in the surface layer(s) (i.e., "surface bond rehybridization" [20,59]) which, in turn, induce elastic distortions in the layers beneath. The surface geometry is that expected for III-V compounds: the anion exhibits a distorted p^3 conformation and the cation a nearly planar sp^2 conformation. Second and deeper layer distortions occur, but are much smaller than those.in the top layer and hence are less reliably extracted from the experimental measurements [1]. The central conceptual issue is why are the surface geometries of II-VI compounds so similar to those of III-V compounds [3]? The central experimental issues are (1) how accurately (and reliably) are the individual atomic geometries known and (2) how similar are the (110) surface geometries of the various III-V and II-VI compounds?

In order to address both experimental issues, we begin by defining the concept of a bond-length-conserving top-layer rotation [1,24]. The independent surface structural variables are indicated in Fig. 7 [6]. The top layer can be relaxed in such a way that all bond lengths are held constant so that the only independent structural variable is the tilt angle, ω_1, between the plane of the chains of atoms in the relaxed top layer and the plane of the unrelaxed surface. For normally incident electrons, two regions of locally optimal fits are obtained via this procedure, one in the vicinity of $0 \leq \omega_1 \leq 10°$ and one in the vicinity of $26° \leq \omega_1 \leq 33°$ [1,34-37]. Although considerable controversy arose over the existence of two minima [5,6], it is now recognized that both exist and that the large-ω minimum corresponds to the physically realized structure [1,4-6,34-37]. Further refinement of this minimum utilizing the full range of independent structural variables (Fig. 7) permits an improvement to be achieved in the theoretical description of experimental LEED intensity data [1,4-6,34-37].

Table 1 contains a listing of the three structures obtained by first identifying the two locally optimal bond-length-conserving geometries, optimizing each of them, and selecting the overall "best fit" between the resulting structures [1]. The absolute accuracy of distances normal to the surface is about ±0.1 Å and of those parallel to the surface is about ±0.4 Å, although much higher precision can be achieved in the analysis of a particular data set from a given crystal. Within the overall accuracy, however, all of the extant structure analysis agree [1,5,6,35,37] and are compatible with a bond-length-conserving top-layer rotation corresponding to $\omega_1 = 29° \pm 3°$ for all zincblende (110) surfaces, even though there is considerable scatter within this range [1,15] and the magnitude of the atomic displacements parallel to the surface is still in question

Table 1 Optimal bond-rotated and multi-parameter best-fit structures for the (110) surfaces of naturally-occurring zincblende-structure compound semiconductors as determined from LEED intensity analysis. The structural parameters are defined in Fig. 7. The structures labelled "low ω_1" and "high ω_1" are the local optimal structures obtained subject to the constraint, that only bond-length conserving top-layer rotations are considered. "Best fit" structures are the overall optimal structures. Values of the X-Ray (R_X) and integrated intensity (R_I) R factors indicate the overall quality of the fit to the measured LEED intensities for the various locally optimal structures. (Adapted from Duke [1].)

Compound	Structure	Layer	Anion (Å)	Cation (Å)	$\Delta_{1,\perp}$ (Å)	$\Delta_{2,\perp}$ (Å)	$d_{12,\perp}$ (Å)	$d_{23,\perp}$ (Å)	$\Delta_{1,y}$ (Å)	$d_{12,y}$ (Å)	ω_1 (deg)	R_X	R_I
AlP	Low ω_1	1	↑0.06	↓0.06	0.12	0	1.86	1.93	4.09	2.81	5.0	0.28	0.22
	High ω_1	1	↑0.19	↓0.44	0.63	0	1.49	1.93	4.24	3.20	27.5	0.26	0.04
	Best Fit	1	↑0.06	↓0.57									
					0.63	−0.07	1.33	1.96	4.11	2.96	25.2	0.19	0.02
		2	↓0.03	↑0.04									
GaP	Low ω_1	1	↑0.03	↓0.03	0.06	0	1.90	1.93	4.09	2.77	2.5	0.29	0.18
	High ω_1	1	↑0.19	↓0.44	0.63	0	1.49	1.93	4.24	3.20	27.5	0.25	0.07
	Best Fit	1	↑0.19	↓0.44	0.63	0	1.39	1.93	4.24	3.20	27.5	0.22	0.07
GaAs	Low ω_1	1	↑0.08	↓0.09	0.11	0	1.90	2.00	4.25	2.95	7.0	0.20	0.35
	High ω_1	1	↑0.20	↓0.49	0.69	0	1.51	2.00	4.42	3.34	29.0	0.21	0.13
	Best Fit	1	↑0.16	↓0.53									
					0.69	−0.06	1.44	2.02	4.52	3.34	31.1	0.17	0.12
		2	↓0.03	↑0.03									
GaSb	Low ω_1	1	↑0.11	↓0.16	0.27	0	2.01	2.16	4.61	3.26	10.0	0.28	0.14
	High ω_1	1	↑0.22	↓0.55	0.77	0	1.62	2.16	4.79	3.63	30.0	0.24	0.05
	Best Fit	1	↑0.22	↓0.55	0.77	0	1.62	2.16	4.79	3.63	30.0	0.24	0.05
InP	Low ω_1	1	↑0.06	↓0.01	0.13	0	2.01	2.08	4.41	3.03	5.0	0.31	0.06
	High ω_1	1	↑0.21	↓0.52	0.73	0	1.55	2.08	4.60	3.48	30.0	0.19	0.03
	Best Fit	1	↑0.21	↓0.52	0.73	0	1.55	2.08	4.60	3.38	30.0	0.18	0.03
InAs	High ω_1	1	↑0.22	↓0.56	0.78	0	1.57	2.13	4.74	3.60	31.0	0.28	0.04
	Best Fit	1	↑0.22	↓0.56									
					0.78	−0.15	1.50	2.21	4.99	3.60	36.5	0.23	0.08
		2	↓0.07	↑0,08									
InSb	Low ω_1	1	↑0.10	↓0.13	0.23	0	2.16	2.29	4.87	3.41	8.0	0.22	0.45
	High ω_1	1	↑0.23	↓0.65	0.88	0	1.64	2.29	5.12	3.89	33.0	0.26	0.03
	Best Fit	1	↑0.18	↓0.70									
					0.88	−0.10	1.54	2.34	5.32	3.89	37.3	0.21	0.01
		2	↑0.05	↓0.05									
ZnS	Low ω_1	1	↑0.03	↓0.03	0.06	0	1.88	1.91	4.06	2.75	2.5	0.21	0.23
	High ω_1	1	↑0.18	↓0.41	0.59	0	1.53	1.91	4.19	3.15	26.0	0.22	0.10
	Best Fit	1	↑0.08	↓0.51	0.59	0	1.40	1.91	4.30	3.15	28.0	0.18	0.09
ZnSe	Low ω_1	1	↑0.05	↓0.05	0.10	0	1.95	2.00	4.25	2.91	4.0	0.23	0.26
	High ω_1	1	↑0.20	↓0.49	0.69	0	1.52	2.00	4.43	3.35	29.0	0.31	0.13
	Best Fit	1	↑0.10	↓0.59	0.69	0	1.42	2.00	4.93	3.55	42.9	0.24	0.16
ZnTe	Low ω_1	1	↑0.13	↓0.20	0.33	0	1.96	2.15	4.60	3.29	12.5	0.30	0.12
	High ω_1	1	↑0.21	↓0.51	0.72	0	1.64	2.15	4.75	3.59	28.5	0.28	0.05
	Best Fit	1	↑0.16	↓0.56									
					0.72	−0.05	1.58	2.18	4.68	3.52	26.9	0.25	0.04
		2	↓0.02	↑0.03									
CdTe	Low ω_1	1	↑0.08	↓0.09	0.17	0	2.20	2.29	4.81	3.36	6.0	0.30	0.23
	High ω_1	1	↑0.23	↓0.58	0.81	0	1.71	2.29	5.08	3.84	30.0	0.26	0.03
	Best Fit	1	↑0.18	↓0.63									
					0.81	−0.15	1.59	2.31	5.18	3.84	31.9	0.22	0.02
		2	↓0.01	↑0.08									

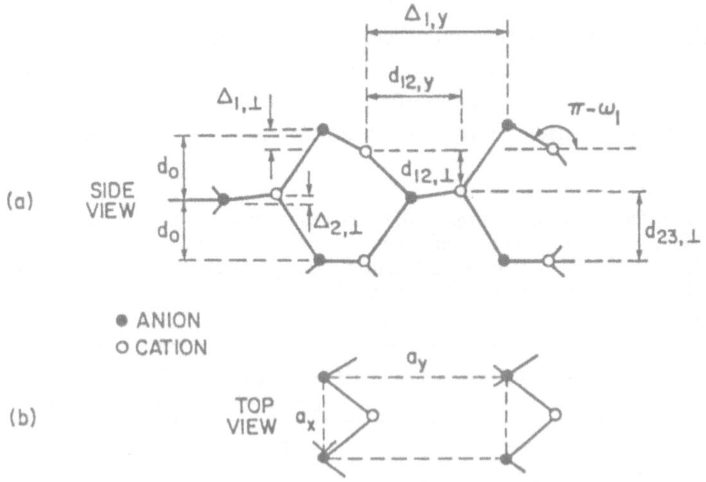

Figure 7 Panel (a): Schematic indication of the surface atomic geometry for the (110) surface of zincblende structure compound semiconductors. The layer spacing is $d_0 = a_0/2\sqrt{2}$ where a_0 is the bulk lattice constant. Panel (b): surface unit cell. Surface unit mesh parameters are $a_y = a_0$ and $a_x = a_0/\sqrt{2}$. (After Duke and Paton [6].)

[6,35,37]. A detailed review of the uncertainties inherent in these analysis has been given by Duke [1] and subsequently refined by Cowell, Prutten, Tear and de Carvalho [34, 35, 37]. Comparison of locally optimal bond-length-conserving rotation structures provides a well-defined quantitative measure of the similarity of the atomic geometries of the (110) surfaces of zincblende-structure compound semiconductors. Confirming our attention to the physically-realized large ω_1 structures, we see from Table 1 that all of these structures occur in the range $26° \leq \omega_1 \leq 33°$. These structures can be refined further by permitting additional structural parameters to vary, but only small additional improvements in the quality of fit to the experimental intensities are achieved [1] and these are sensitive to how the parameters are determined [35, 37]. Therefore the results shown in Table 1 suggest that to first order the atomic geometries of the (110) surfaces of all zincblende structure compound semiconductors are the same and are characterized by $\omega_1 = 29° \pm 2°$ independent of the specific semiconductor material. Consideration of other experimental results expands the uncertainties somewhat to $\omega_1 = 29° \pm 3°$ [7,35] but the basic conclusion remains unaltered. Moreover, isoelectronic materials with widely different ionicities exhibit nearly identical optimal values of ω_1 (e.g., GaAs and ZnSe [60], GaSb and ZnTe [52], GaP and ZnS [53], InSb and CdTe [35,37,61]): a result which has been reconfirmed within the past two years [35-37]. Therefore an important concept to emerge from the zincblende (110) experimental structure determinations is the notion that all III-V and II-VI materials exhibit a "universal"

atomic geometry when their independent structural variables are scaled with the bulk lattice constants [1]. If these variables are constrained to be those associated with a top-layer bond-length-conserving rotation alone, the rotation angle $\omega_1 = 29° \pm 3°$ is obtained. Obviously other structural variations also occur [1,35,37] but they need not contradict the idea of a universal zincblende (110) surface structure for sp^3 bonded materials and, indeed, are compatible with it to within the accuracy of existing data analyses.

Table 2 Studies of the surface structures of the (110) surfaces of zincblende-structure compound semiconductors. Entries in the table are reference numbers. The abbreviations used in the table heading are LEED: low energy electron diffraction, PES: photoelectron spectroscopy, IS: ion scattering, an ESR: electron spin resonance. The GaAs references either are to reviews or to work published since 1986.

Material	LEED	PES	IS	ESR	Theoretical prediction	Surface State Eivenvalues
AlP	62	—	—	—	17,80	79
AlAs	63	—	—	—	17,80	79
AlSb	—	—	—	77,78	81,82	7g
GaP	53,64,65	—	—	77,78	17,80,82,83	86,87,88
GaAs	1,4-6	1	1,7,8,11	—	1,19,45,46	1,45
GaSb	52,66	—	67,68	77	81,83	86,87
InP	1,69,70	72	—	—	19,81,83,84	72,86,89,90
InAs	71	—	68	—	81,83	86,90
InSb	1,35,73	—	—	—	81,83,84	86,90,91
ZnS	53,74	—	—	—	19,42	42
ZnSe	60	12,76,103	—	—	19,43,47,84,85	12,43,47,76,92,93
ZnTe	52,75	—	—	—	84	76
CdTe	1,36-38,61	44,94	—	—	19,44	44

Finally, we provide in the first four columns of Table 2 a list of citations, to the papers in which experimental determinations of the atomic geometries of zincblende (110) surfaces have been reported. In the entry for GaAs(110) only references since those reviewed by Duke [1] are given due to the voluminous nature of this list and its availability in the published literature [1, 2]. The last two columns of Table 2 contain citations to theoretical predictions of these geometries and their associated surface-state eigenvalue spectra. We turn next to discussions of each of these two topics, in turn.

2.2 Atomic Geometry – Theoretical Prediction

The major conceptual issue to be addressed by theoretical models is the interpretation of the similarity of the experimentally-determined (110) surface atomic geometries of III-V and II-VI compound semiconductors. We found earlier that in contradiction with theoretical expectations based on local coordination chemistry [17] and ionicity-structure correlations [18,19,48-51], the similarity (indeed identity to within experimental uncertainties) of the (110) atomic geometries of four pairs of isoelectronic materials (GaSb/ZnTe [52], GaP/ZnS [53], GaAs/ZnSe [60] and InSb/CdTe (35, 37, 61]) with widely different ionicities and small-molecule coordination chemistries has been established. In addition, the existence of a "universal" zincblende (110) surface structure characterized by bond-length-conserving top-layer rotations of $\omega_1 = 29° \pm 3°$ to within current experimental uncertainties [1,15] encompasses the II-VI materials ZnS, ZnSe, ZnTe and CdTe as well as the common III-V compound semiconductors. (See Tables 1 and 2.) An explicit explanation of this result was proposed [1] in 1985; i.e., that the reconstructions were pseudo-Jahn-Teller effects caused by the rehybridization of dangling bond states on the surface anions into back-bonding surface-state bands with the concomitant lowering in energy of these bands providing the driving force for the reconstruction. Such a reconstruction mechanism is highly dependent on the surface topology (i.e., on the fact that activationless bond-length-conserving rotations can occur at zincblende (110) cleavage surfaces) but is less dependent upon the details of the electronic structure. The quantitative validity of this idea could be established for III-V compounds via calculations available by 1985 [80,81,83]. At that time, however, theoretical predictions were available only for ZnSe and ZnTe [84,85] among the II-VI compounds, and the scaling of the predicted structures with bulk lattice constant had not been examined explicitly for these compounds. In light of the importance of extending model calculations to II-VI compounds, one major thrust of recent work has been the extension of spectroscopically parameterized tight-binding total energy models, which had proved highly successful in predicting III-V (110) surface geometries [83-85], to II-VI materials. Specifically, a sp^3s^* model developed for bulk III-V compounds [83,95] is used for ZnSe [43] whereas sp^3 models developed from bulk spectroscopic data are utilized for ZnTe [84], ZnS (42], ZnO [96], CdTe [44], CdSe [97, 98] and CdS (97]. The energy is written as the sum of an electronic part (band-structure energy, E_{bs}) and an elastic part, i.e.,

$$E_{tot} = E_{bs} + \sum_{i<j} (U_1 \epsilon_{ij} + U_2 \epsilon_i^2) \tag{2.1}$$

in which ϵ_{ij} is the fractional change in the bond length away from its bulk crystalline value. E_{bs} is evaluated by summing over the occupied states of the energy spectrum. The parameter U_1 is determined by the equilibrium conditions and U_2 is determined by the bulk elastic modulus, B, of a crystal of volume V i.e.,

$$U_1 = - \left. \frac{\partial E_{bs}}{\partial \epsilon} \right|_{\epsilon=0}, \tag{2.2}$$

$$2U_2 = 9VB - \left. \frac{\partial^2 E_{bs}}{\partial \epsilon^2} \right|_{\epsilon=0} \tag{2.3}$$

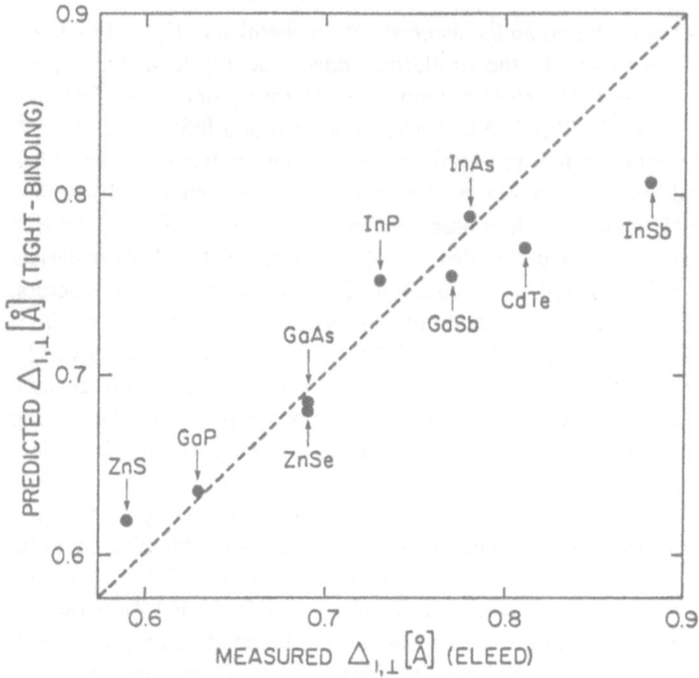

Figure 8 Top-layer shear, $\Delta_{1,\perp}$ predicted by tight-binding total energy minimization, plotted against the value determined by dynamical ELEED data analysis for the (110) surfaces II-VI and III-V compounds crystallizing in the zincblende structure (After Wang et al. [44].)

An eight layer slab is used in most of the calculations to simulate a semi-infinite crystal. The surface atomic geometry is determined by minimizing the total energy in Eq. (2.1). The changes in the interatomic interactions due to the relaxation of the atoms in the top two layers are accounted for by use of the d^{-2} scaling law [99] for the orbital interactions. Details of the parameterizations and calculations may be found in the references cited [41-44,83-85,95-98]. A comparison of the predicted structures with those obtained from elastic low-energy electron diffraction (ELEED) intensity analysis is shown in Fig. 8 for the case of $\Delta_{1,\perp}$ (see Fig.7) which is the structural parameter most accurately extracted from experimental ELEED intensities. Evidently, these models yield a quantitative prediction of the vertical displacements within the top layer associated with the experimentally determined surface atomic geometries for the (110) surfaces of zincblende structure compound semiconductors. In addition, the models reveal explicitly that the rehybridization induced lowering of the anion dangling-bond surface states provides the major driving force for the reconstruction and hence link the observed "universality" of the reconstructed surface atomic geometries with the surface-state rehybridization mechanism for the reconstructions.

The utility of tight-binding total-energy models flows from their being sufficiently

Table 3 Atomic displacements in Å from unrelaxed positions for the (110) surface of GaAs, compared to previous calculations and to structures determined by ELEED intensity anlaysis. The first two lines give vertical displacements of the first-layer As and Ga. $\Delta_{1,\perp}$, $\Delta_{2,\perp}$ $d_{12,\perp}$, $\Delta_{1,y}$, and ω_1 are defined in Fig. 7. $\omega_1 = \sin^{-1}(4\Delta_{1,\perp}/a_0)$. (After Qian et al. [46].)

	Theory				LEED			
	Present	Chadi Ref. 84	Maihiot et al. Ref. 83	Ferraz and Srivastava Ref. 45	Meyer et al. Ref. 101	Duke et al. Ref. 4	Tong et al. Ref. 102	Puga et al. Ref. 5
As	0.143↑	0.186↑	0.176↑	0.2351↑	0.144↑	0.159↑	0.176↑	0.193↑
Ga	0.442↓	0.459↓	0.507↓	0.510↓	0.506↓	0.527↓	0.510↓	0.515↓
$\Delta_{1,\perp}$	0.58	0.65	0.68	0.75	0.65	0.686	0.686	0.708
$\Delta_{2,\perp}$	−0.07	−0.13	−0.20	−0.035	−0.12	−0.06	−0.03	−0.06
$d_{12,\perp}$	1.44	1.47	1.35	1.46	1.43	1.44	1.47	1.45
$\Delta_{1,y}$	4.39	4.40	4.45	4.45	4.40	4.52	4.36	4.43
$d_{12,y}$	3.18	3.17	3.18	3.36	3.31	3.34	3.17	3.18
ω_1	27.4°	27.3°	29.7°	31.6°	27.3°	31.1°	28.0°	30.1°

simple and transparent that issues like the mechanism of (110) surface structure universality can be explored quantitatively yet explicitly. Their spectroscopic parametrization to describe bulk optical and photoemission spectra permits them to be used to extrapolate from bulk phenomena to surface phenomena reasonably reliably, as evidenced from Fig. 7 and the detailed results on which it is based [42-44,83]. Nevertheless, their lack of an explicit treatment of coulomb interactions renders their quantitative predictions suspect, especially for highly ionic materials (e.g., ZnO, ZnS and ZnSe). Thus, it is important to calibrate such models by comparison with more accurate calculations. For GaAs(110) two such calculations have appeared recently. Qian et al. [46] have reported an extensive *ab initio* pseudopotential analysis in which the absolute cleavage energy was evaluated with emphasis on achieving a completely converged result which, moreover, reproduced the observed (100) value ($E_c = 1.21 \pm 0.21$ eV/cell) to better than 1 %. They compared their structural predictions to those of two tight-binding calculations [83,84], four ELEED intensity analyses [4,5,101,102], and a recent empirical pseudopotential calculation for GaAs(110) by Ferraz and Srivastava [45] with the results shown in Table 3. Recognizing that the absolute accuracy of $\Delta_{1,\perp}$, $\Delta_{2,\perp}$, and $d_{12,\perp}$ is uncertain to approximately 0.1 Å and that of $\Delta_{1,y}$ and $d_{12,y}$ is uncertain to approximately 0.4 Å from the ELEED intensity analysis methodology [1], we see from Table 3 that the theoretical predictions are compatible with the experimental structures but that the disagreements among the predictions are much larger than those among the different independent ELEED intensity analyses. Table 3 represents the current state-of-the-art in theoretical calculations of zincblende (110) surface structures, revealing that the predictions of purportedly accurate "first principles" calculations can differ from each other by amounts greater than or comparable to the experimental uncertainties in the ELEED analyses (e.g., 0.1 Å for $\Delta_{1,\perp}$. Therefore for GaAs(110) small (e.g., 0.1 Å) variations in structural parameters neither can be established experimentally nor predicted theoretically with reliability at the present time. This conclusion also pertains to ZnSe(110) which is the other surface structure predicted by both tight-binding [43] and empirical pseudopotential [47] methods during the past three years.

The final recent development in the theory of zincblende (110) surface geometries is the appearance of a series of papers by Tsai and coworkers [19, 48, 49] resurrecting the idea [18, 23, 50, 51] that the character of the surface reconstruction should depend systematically on the ionicity of the bulk material. The issue is not whether a dependence of surface structure on ionicity exists or not, but rather is the magnitude of this dependence and the specification of the structural parameters in which it manifests itself. Tables 1 and 3 illustrate well the difficulties inherent in isolating such an effect at the current state-of-the-art in both experimental surface structure determination and theoretical total-energy calculation. Both tables reveal that only the major features of the surface structure may be established reliably at the present time. Displacements parallel to the surface in the top layer and all displacements in the second and deeper layers are sufficiently imprecisely established that refinements to the top-layer bond-length-conserving-rotation model are not known with adequate accuracy to establish trends with ionicity. Although the predictions of Tsai and coworkers are grossly at variance with the available experimental results [19,48], they argue that the trend which they predict, i.e., that ω_1 decreases with increasing ionicity, is valid nevertheless. A more convincing case for this type of argument was advanced in 1981 by Swarts, McGill and Goddard [80] who calculated $\omega_1 = 20.6° \pm 1.2°$ for the nitrides of B, Al and Ga but $\omega_1 = 27.4° \pm 0.5°$ for their arsenides and phosphides. This prediction has not been confirmed experimentally because zincblende structure crystals of the nitrides are not available. It certainly is possible that within the uncertainties inherent in the $\omega_1 = 29° \pm 3°$ top-layer, bond-length-conserving-rotation model a systematic variation of the value of ω_1 with ionicity occurs which would manifest itself experimentally if crystals of zincblende-structure nitrides could be grown with adequate perfection to enable ELEED or ion-scattering experiments. The demonstration of such trends, requires, however, either improved accuracy of the surface structure methodologies or the availability of crystals of materials (e.g., zincblende-structure nitrides or chlorides) whose bonding is sufficiently dissimilar from the arsenides, phosphides, antimonides, sulfides, selenides and tellurides to reveal a trend within the present state-of-the-art.

2.3 Electronic Surface States and Resonances

As noted in the preceding section, electronic surface states play a special role in the theory of zincblende (110) atomic geometries because it is the lowering of a nearly-mid-gap anion dangling-bond surface state, with its concomitant rehybridization into a back and surface bonding state which provides the common driving force for an

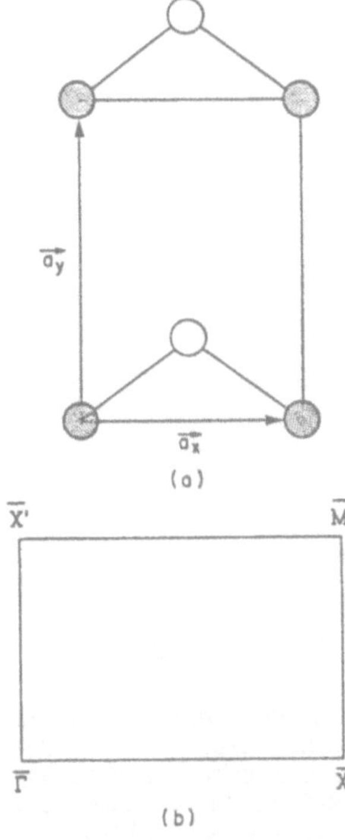

Figure 9
Schematic indication of the surface unit mesh [panel (a)] and associated two-dimensional Brillouin zone [panel (b)] of the (110) surface of a zincblende structure compound semiconductor with the surface atomic geometry specified as indicated in Fig. 7.

approximately "universal" surface reconstruction. Thus, the confirmation of this mechanism of the surface reconstruction requires the further prediction of the surface state eigenvalue spectra and their experimental confirmation, e.g., by Angle Resolved Photoemission Spectroscopy (ARPES). Using the tight-binding models, the surface electronic states and surface resonances are identified by examining the calculated symmetry-resolved charge densities and locating eigenfunctions which are localized within the top layers. The results are presented as surface-state energies along line segments in the two-dimensional surface Brillouin zone. For zincblende (110) the surface unit mesh and Brillouin zone are indicated in Fig. 9. The Brillouin zone is that area of momentum space for momenta parallel to the surface, $k_{||}$, which contains all of the independent values of $k_{||}$. Typically, energy eigenvalues versus $k_{||}$, are shown along the lines in momentum space which bound the Brillouin zone.

Predicted eigenvalues for CdTe(110) are shown in this fashion in Fig.10 in which they are compared with values extracted from ARPES measurements [44]. Shaded areas in the figure indicated the projection of the bulk energy bands on the surface Brillouin zone. Solid lines indicate eigenvalues associated with true (i.e., localized) surface states whereas dashed lines indicate those of surface resonances. The surface relaxation is driven by the lowering of the energies of the highest-energy occupied (anion derived) surface states ("S_1 ") from the dot-dashed line resulting for the un-

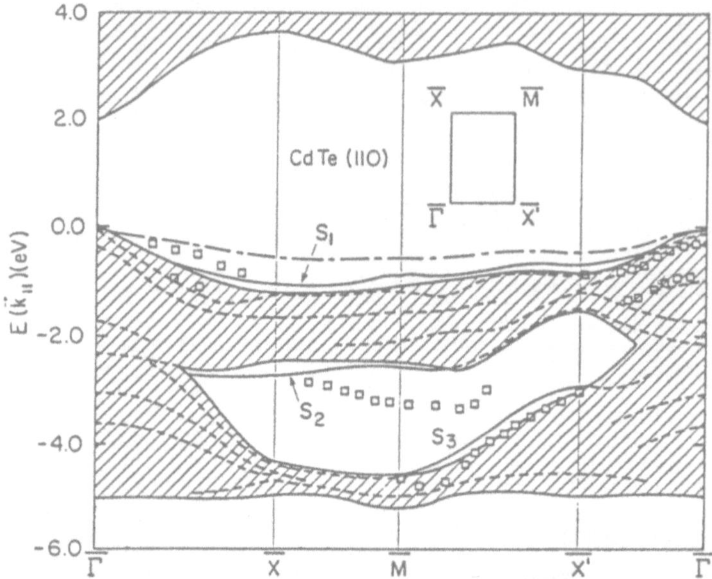

Figure 10 Band structure of the (110) surface of CdTe. The surface Brillouin zone in indicated in the inset. The squares designate surface-state energies extracted from experimental angle-resolved photoemission measurements. The calculated surface-state and surface-resonance eigenvalues are shown by solid and dashed lines, respectively. The dot-dashed line designates the highest-energy occupied surface state S_1 of the ideal (110) face. (After Wang et al. [44].)

relaxed surface to the values indicated in the figure. The experimental surface-state energies extracted from ARPES results (designated by squares in the figure) are consistent with the predicted surface-state eigenvalues following the lowering, and hence with the proposed interpretation of the mechanism of surface relaxation. This proposition was originally proposed and confirmed for GaAs(110) [1], and has been verified qualitatively for ZnSe(110) as well [12, 43, 47,103,104].

Inspection of Fig.10 further reveals that the model predictions are in qualitative, but not generally quantitative agreement with the ARPES measured surface-state eigenvalue spectra for CdTe(110). This situation also prevails for GaAs, InP and ZnSe, the other three materials whose ARPES have been examined in detail [1]. Moreover, for GaAs(110) several independent sets of ARPES measurements of surface-state eigenvalues have been given which are not fully compatible one with another, due possibly to varying methods of surface preparation, different photoemission geometries and/or sources, and non-identical data analysis procedures [1,2,12]. The theoretical calculations exhibit their uncertainties as well, for example those associated with the fact that the electronic relaxation around a photogenerated hole is different for holes localized at the surfaces in surface states than for those in bulk states. The existence of these uncertainties in the detailed acquisition and interpretation of valence-electron surface-state eigenvalue spectra from photoemission data renders currently available spectra of limited utility in clarifying the quantitative features of various models of surface electronic structure, although their qualitative features confirm the pseudo-Jahn-Teller surface-state lowering mechanism of the zincblende (110) surface reconstruction [44].

2.4 Vibrational Surface States and Resonances

Since the same forces which drive the reconstruction of zincblende (110) surfaces govern the atomic vibrations of the surface atoms, it is natural to expect that studies of vibrational surface states (i.e., vibrational modes localized near the surface) would illuminate the nature of these forces. Indeed, one of the major new experimental results reported for zincblende (110) surfaces in the past few years is the examination of the surface atom dynamics of GaAs(110) by Helium Atom Scattering (HAS) [39,40]. Conceptually, we visualize the surface atoms as moving on a potential energy surface (PES) whose minima define the reconstructed surface geometry and whose curvatures in the vicinity of these minima define the vibrational motions of the atoms. The tight-binding total-energy model described in Sec. 2.2 describes the PES for the various III-V and II-VI materials and, as we found in conjunction with Fig. 8, predicts the minimum-energy positions on the PES for the (110) surface quite well. Specifically, since the surface reconstruction is approximately a bond-length-conserving rotation, the total energy exhibits a valley along the rotational trajectory. The minimum of this value determines the relaxed surface atomic geometry. Quantized vibrations (rotations) along the trajectory around the energy minimum constitute the rotational surface phonons. The phonon frequencies, ω, can be determined from the total energy difference between the fully relaxed surface and the one slightly rotated from the equilibrium geometry via use of an expression obtained from the virial theorem, i.e.,

$$\tfrac{1}{2}M_a\omega^2u_a^2 + \tfrac{1}{2}M_c\omega^2u_c^2 = \Delta E_{tot} \ , \tag{2.4}$$

where M_a, u_a and M_c, u_c are the mass and displacement of the top layer anion and cation, respectively. ΔE_{tot} is the total energy change per surface unit cell; u_a and u_c are calculated from the rotated surface geometry. Hence, if we know the atomic displacements, u_i, we can calculate the vibration frequencies, ω. These displacements can be found by using the tight-binding total-energy model to calculate the effective atom-atom forces and then inserting these forces into a dynamical calculation of the eigenfunctions and eigenvalues of the atomic displacements from the reconstructed geometry: a procedure that has been performed for Si [105,106] and its (100) [106,107] and (111) [108,109] surfaces. Alternatively, for the zincblende (110) surfaces one can note that lowest-energy motions of the surface anions and cations must correspond roughly to bond-length-conserving rotations about the reconstructed geometry because this is the only motion possible for an incompressible lattice (i.e., large values of U_2 in Eq. (2.1)). For such rotations the u_i in Eq. (2.4) can be evaluated from geometrical constraints alone. Hence the value of ω for such a surface rotational mode can be obtained from Eq. (2.4) because E_{tot} is given by the tight-binding total energy model. This analysis has been performed by Wang and Duke [41] for the (110) surfaces of GaP, GaAs, GaSb, InP, InAs, InSb, ZnS, ZnSe and CdTe. They found $\hbar\omega = 10.7$ meV for GaAs(110) using the same model [83] that predicted the measured reconstructed atomic geometry to within experimental uncertainties (See Table 3).

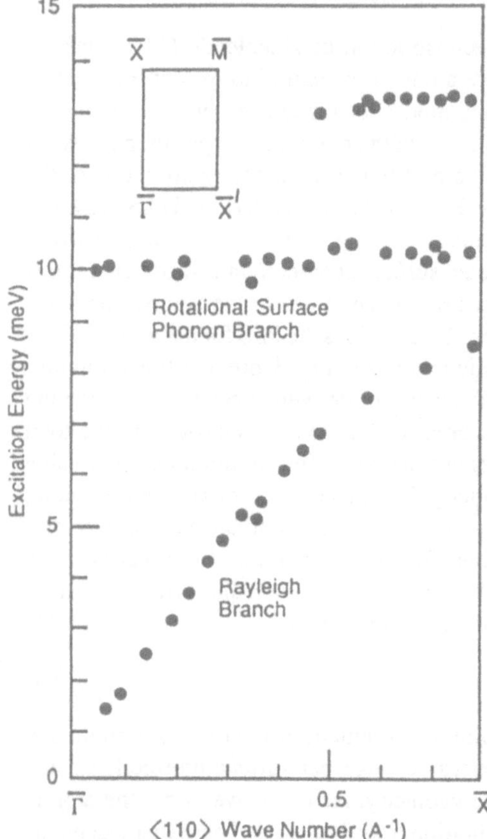

Figure 11
Measured surface phonon dispersion curves along the $\overline{\Gamma} - \overline{X}$ line in the surface Brillouin zone. The surface temperature is 300 K. The surface Brillouin zone is shown in the inset. The rotational surface phonon energies are predicted by Wang and Duke [41] are $\hbar\omega(\overline{\Gamma}) = 10.7$ meV and $\hbar\omega(\overline{X}) = 12.6$ meV. (After Duke and Wang [28].)

In 1987, Harten and Toennies [39] reported the occurrence of a new "optical " branch of surface phonons with $\hbar\omega = 10$ meV for GaAs(110) along the $\overline{\Gamma} - \overline{X}$ line in the surface Brillouin zone. Their results, obtained from an analysis of inelastic HAS, are shown in Fig.11. In addition to the acoustical ("Rayleigh") branch of the surface vibrational spectrum, there are two flat ("optical") branches, presumably associated with motions of the top-layer Ga and As species relative to the substrate which are accompanied with little or no bond stretching. The lower of these corresponds well with the estimates of Wang and Duke [41] with regard to both energetics and selection rules [28].

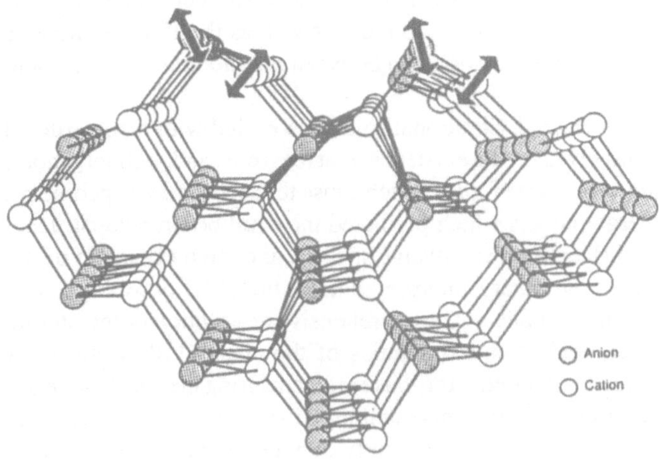

Figure 12 Schematic indication of the rotational surface vibrations on the (110) surfaces of zincblende structure compound semiconduc-tors. The chains lie along the < 110> direction in the surface and the normal mode displacements occur in the plane perpendicular to this direction. Motions of atoms below the top layer are much smaller than those of the top-layer atoms. (After Duke and Wang [28].)

The actual lattice dynamics of GaAs(110) are, of course, considerably more complex than suggested by the incompressible-bond limit. An indication of the qualitative features of these dynamics may be found from inspection of the results for the Si(111)-(2 × 1) "Pandey π-bonded chain model" [108-110]. The top two layers of this surface structure resemble those of GaAs(110), although the detailed electronic origin of the reconstruction is quite different in the two cases. Rotational surface modes are found for Si(111)-(2 × 1), although they occur at higher energies ($\hbar\omega(\overline{\Gamma}) \approx 30$ meV) than the surface chain "bounce" modes ($\hbar\omega(\overline{\Gamma}) \approx 10$ meV) which correspond to the Ga and As vibrating in phase (i.e., an "acoustical" mode rather than an out-of-phase ("optical") mode) normal to the surface [108,109]. Near the edge of the zone the bounce mode becomes a vibration of only one sublattice. This bounce-to-single-sublattice mode on Si(111)-(2 × 1) is roughly analogous to the 10 meV surface rotational mode of GaAs(110), and reveals the complexities relative to the incompressible bond limit that a thorough treatment of the lattice dynamics can introduce. Indeed, the two model calculations in the literature [109,110] do not agree among themselves as to the exact lattice dynamics. Nevertheless, all treatments agree that for $\hbar\omega < 30$ meV the surface modes consist predominately of surface-chain atomic motions relative to the substrate so that these modes may be visualized schematically as shown in Fig.12 for zincblende

(110). Indeed, the results for the Si(111)-(2×1) Pandey π-bonded chain model suggests that the mode evident in Fig. 11 at $\hbar\omega(\overline{\Gamma}) \approx 13$ meV as well as the 10 meV mode is of this character, but with a different relationship between the motions of the As and Ga atoms than the 10 meV mode.

In summary, the study of localized vibrational states associated with reconstructed zincblende (110) surfaces has revealed the existence of at least one, and probably more, surface states characteristic of the same forces which cause the reconstructed geometry. The tight-binding total energy models which predicted the equilibrium reconstructed geometries, the electronic surface state eigenvalue spectra, and the mechanism of the reconstruction also afford a preliminary interpretation of the HAS observations of surface vibrational modes. Thus, a coherent, comprehensive description of the atomic structure, electronic structure and atomic dynamics of the clean (110) surfaces of zincblende structure compound semiconductors has emerged during the past few years. At the present time this description is still only semiquantitative, but it seems capable of providing a unified interpretation of a wide variety of experimental measurements to within their inherent uncertainties.

2.5 Adsorption on Zincblende (110) Surfaces

Although there is an enormous literature on adsorbates on zincblende (110) surfaces, relatively few of these systems exhibit epitaxical growth. Fewer still have been the subjects of quantitative surface structure analyses. The importance of such analyses from the perspective of the present review is their revelation that chemical reactions with adsorbates usually modify the reconstruction of the underlying substrate. Moreover, such modifications are profoundly process-dependent; i.e., the resulting structures are dependent on the dosing and thermal history of the composite system as well as on the morphology of the initial substrate. These structures are generally metastable and can be modified by further processing. Hence for thoroughly-studied systems, e.g., GaAs(110)-p(1 × 1)-Sb(IML) [1,20,27], a discussion of the resulting atomic geometries, growth habits, and surface morphologies can become quite complex.

Given this situation, our treatment of adsorbate systems will be brief and focussed upon illuminating the novel types of surface chemical bonding resulting from the combination of the chemical properties of the adsorbate and the template effect of the substrate. Citations will be predominately to reviews and post 1985 literature, from which the interested reader can gain access to a more complete description of the topic.

Group V Adsorption: Epitaxically Constrained Adsorbate Bonding

A central concept in semiconductor surface chemistry is that geometrical constraints imposed by epitaxical growth induce new types of adsorbate-substrate bonds [20]. One of the most extensively studied semiconductor adsorbate systems is the p(1 × 1) Sb saturated monolayer structure on the (110) surfaces of GaAs [111-124] and InP

[124-130]. More recently Bi also has been shown to form saturated monolayers on GaAs(110) [54-58], and has attracted ongoing work to illuminate its similiarities to and differences with the case of Sb [131-133]. Both systems exhibit epitaxically-constrained adsorbate bonding [20,86,134].

Given the chemistry of Sb (or Bi) and of a reconstructed III-V (110) substrate, one can envisage several plausible bonding configurations of an Sb monolayer as indicated in Fig. 13. Historically, the epitaxical continued-layer structure (ECLS) and p^3 structure, shown in panels (a) and (b) respectively of Fig. 13, were proposed [135] to interpret photoemission spectra which did not, however, suffice to distinguish between them. This distinction was achieved subsequently via an ELEED intensity analysis [112] in which the overlapping chain-structure (OCS); shown in panel (d) of Fig.13, also was proposed as a third chemically viable possible overlayer atomic geometry but was found, together with model, to give inadequate descriptions of the ELEED intensities. More recent tunnelling measurements [123] proved compatible with both the ECLS and p^3 geometries. Hence, they motivated a renewed examination of this issue which, in turn, revealed that the p^3 model was unstable to relaxation into the epitaxical on-top structure (EOTS) shown in panel (c) of Fig.13 [136]. This model has not been tested against the measured ELEED intensities for Sb on GaAs(110), but has been found to give an inadequate description of these intensities for GaAs(110)-p(1×1)-Bi(IML) [133]. Hence the ECLS seems to have been established fairly uniquely for saturated p(1×1) monolayers of both Bi and Sb on GaAs(110) [27,58,112,123,133].

In the ECLS, the Sb (or Bi) occupy surface sites in zig-zag chains roughly where the top layer anions and cations would have been. The Sb species are larger than Ga and As, however, so that in order to accommodate the large Sb-Sb bond length (2.8 Å), the angle between neighboring Sb species closes to 91° from its value of approximately 109° in GaAs. In this structure, three of the Sb electrons participate in bonds and the remaining two generate a *sp* hybrid lone pair charge density which is analogous to the charge density for bonding to the substrate species but pointing out of the surface. Such bonding is similar to that of S_8 and Se_8. Thus, the valences of both surface Sb species are fully saturated at monolayer coverage so that the main driving force for the relaxation of the Ga,As(110) substrate is eliminated by the Sb-substrate bonding. Indeed, both the Sb chains and the GaAs(110) substrate are nearly unrelaxed [112]. The Sb bonded to the substrate As is lowered by 0.1 Å relative to that bonded to the substrate Ga and the uppermost As is relaxed downward by a corresponding 0.1 Å[112]. These are small distortions relative to the relaxations on a clean GaAs(110) surface, however, and hence the removal of the substrate reconstruction upon adsorption of the Sb overlayer reveals dramatically the profound influence of a saturated valence top-layer structure on the corresponding atomic geometry. The corresponding ELEED structure analysis for Bi reveals analogous results [58], so we confine our subsequent discussion to Sb. Unusual features of the Sb -substrate bonding are the p^2 local bonding conformation within the Sb chain (i.e., approximately 90° Sb-Sb-Sb bond angles) and distorted sp^3 local Sb conformation (approximately 105° bond angles) relative to the substrate. Since there are not enough independent *p* orbitals to support simultaneously

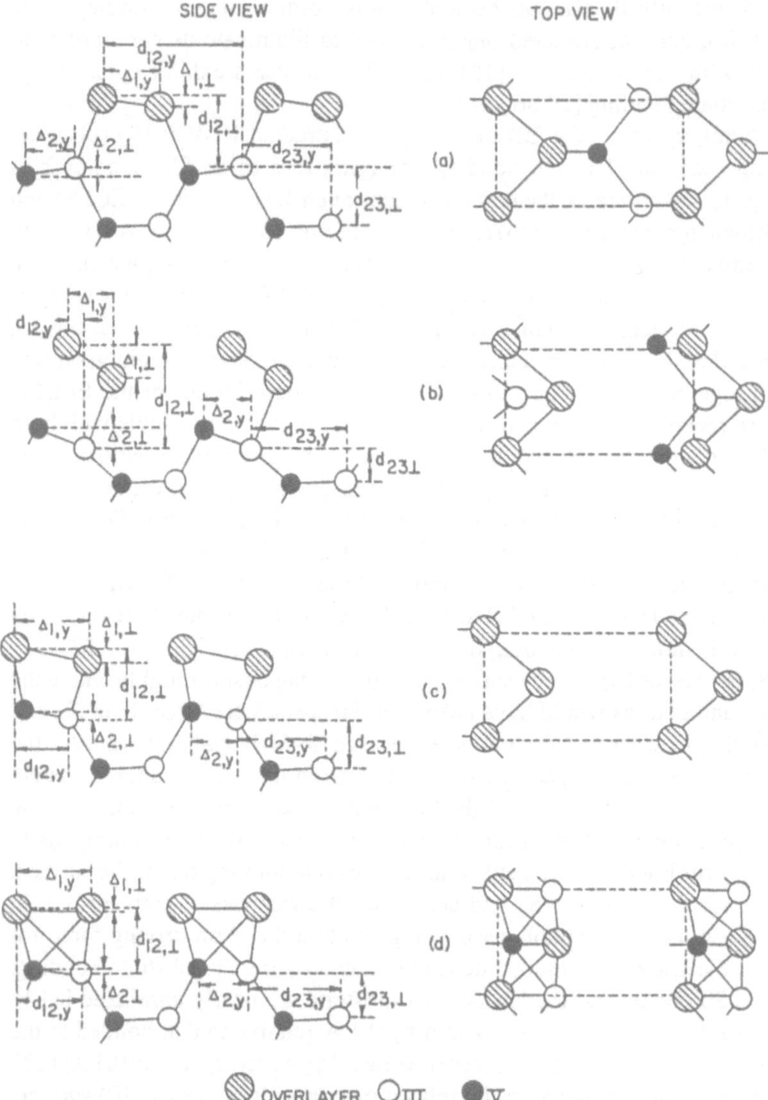

Figure 13 Schematic indication of plausible atomic geometries and associated structural variables of column V monolayers (e.g., Sb, Bi) adsorbed on a III-V (110) substrate. Panel (a): Epitaxical "continued layer" structure (ECLS) in which the column *V* elements occupy sites analogous to those of the clean (110) surface. Panel (b): p^3 structure proposed by Skeath et al. [135]. Panel (c): Epitaxical "on top" structure (EOTS) which is the predicted [136] stable variant of the p^3 structure. Panel (d): Epitaxical overlapping chain structure (EOCS) proposal [112,136] to maximize the overlap charge density between the monolayer and the substrate. (After LaFemina et al. [136].)

both types of local valence, the nature of the bonding is unlike that in either small molecules or the bulk tetrahedrally coordinated solids. Detailed studies [86,134,137] revealed that the Sb-substrate bond is appropriately regarded as a hybrid between π orbitals of a p^2 bonded Sb chain and sp^3 dangling-bond orbitals of an undistorted GaAs substrate. The occurrence of this bond is driven by the combination of the large size of the Sb relative to the GaAs substrate, the saturated valence of the Sb monolayer, and the occurrence of epitaxical ordering of the monolayer relative to the substrate. Thus, the π-sp^3 hybrid bond is a unique type of surface bonding associated with the constraint of epitaxical monolayer growth on the compound semiconductor substrate. Direct confirmation of this type of bonding has been given during the past few years by virtue of direct observations of the surface-state bands characteristic of the bonding charge density via angle resolved photoemission spectroscopy for p(1×1) Sb monolayers on GaAs [115-117,119], GaP [117], and InP [125]. These states are the analogs for the Sb saturated monolayer systems of the back bonding rehybridized surface states which drive the surface rehybridization on the clean cleavage faces of tetrahedrally coordinated compound semiconductors [20]. A recent theoretical analysis [136] indicates, however, that the EOTS (panel (c) of Fig.13) is as compatible with existing photoemission measurements as the ECLS (panel (a) of Fig.13). Therefore, at this time-photoemission experiments do not discriminate definitively between these two classes of structures [136].

Aluminum Adsorption: Process-Dependent Bonding

Al on GaAs(110) is by far the most extensively studied reactive chemisorption system for tetrahedrally coordinated compound semiconductors. Work performed prior to the end of 1983 has been reviewed by Zunger [138] from a theoretical perspective as well as by Skeath etal.[139] and by Bonapace et al. [140] from an experimental one. Herein we review briefly the highlights of studies of the growth habits and atomic geometries of Al on GaAs(110) as an example of a system in which these geometries depend directly on the processing of the deposited overlayer(s).

At room temperature, the deposition of Al on GaAs(110) is thought to proceed via four steps with increasing coverage. At very low coverages, $\Theta \ll 0.1$ ML, the aluminum exhibits mobile chemisorption, probably in two-fold sites [141-144]. As the coverage increases, Al "clusters" form which themselves are mobile (138,139,145-147]. Both of these processes compete with Al-Ga place exchange which is hindered by a high kinetic activation barrier but which can occur at surface imperfections for low coverages. As the coverage is increased further (to $\Theta \approx 1$ ML), the energy gained by the Al coalescing into clusters presumably provides enough energy to surmount this activation barrier locally and Al-Ga place exchange begins on a more extensive scale, resulting in free Ga appearing at the interface. At still higher coverages ($\Theta > 1$ ML), the Al clusters coalesce to form islands which themselves merge to form an Al overlayer (contaminated with Ga) because the "AlAs" interface buffer layer acts as a diffusion barrier which reduces the Al-Ga place exchange in the substrate [63]. At lower substrate temperatures (e.g., $T \approx 100$ K) this sequence of steps is altered by virtue of the lower surface diffusion of the adsorbed Al leading to more homogeneous Al overlayers which distort the underlying atomic geometry of the substrate. Upon warming to room temperature at intermediate coverages ($1ML < \Theta < 2ML$) the surface reverts to its room-termperature deposition form of mobile islands surrounded by patches of relaxed bare GaAs(110), Annealing to 450 °C produces Al-Ga place exchange and hence, the formation of a few atomic layers of Al-rich $Al_xGa_{1-x}As$ [63,148,149]. The exact composition and structure of these layers depends on the initial Al coverage.

LEED studies of the morphology of Al adsorbed on GaAs(110) have been performed for both room-temperature [63,139,148,149] and low-temperature [140] deposition. At room temperature the Al exhibits Volmer-Weber [150,151] growth forming clusters at low coverages which with increasing coverage grow and coalesce to form (three-dimensional) islands. Even when the surface is nearly completely covered (10 ML < Θ < 20 ML) all ordered portions produce the relaxed GaAs(110) clean-surface LEED intensities which could be characteristic either of free patches of GaAs(110) or of a disordered Al overlayer on a weakly-perturbed GaAs(110) substrate. The coverage dependence of the intensities of substrate Ga and As Auger electron lines suggests the former rather than the latter, possibly accompanied by interdiffusion at some regions of the island-substrate interfaces. In contrast, low-temperature (100 K) deposition yields a much more homogeneous growth at low coverages, characterized by Al island growth on a more-or-less uniform interfacial Al overlayer [140]. At Θ < 2 ML a (1 × 1) LEED intensity pattern is observed, but the intensities are characteristic of unrelaxed GaAs(110) indicating either that the interfacial overlayer has restored the GaAs(110) surface atomic geometry to its truncated bulk value, or that the interfacial layer is so disordered by the Al adsorption that only deeper layers of the substrate contribute to the LEED intensities. These truncated-bulk GaAs(110) LEED intensities are observed for 2 ML < Θ < 15 ML. For Θ < 25 ML only the LEED pattern of the epitaxical Al islands is observed.

Quantitative ELEED intensity analyses have been given for the ordered GaAs(110)-p(1 × 1)-Al(Θ) structures prepared by deposition of a nominal coverage of Al followed by vacuum thermal annealing [63,148,149]. Examination of the low-coverage (Θ < 1 ML) structures focused upon evaluating the numerous proposed models of chemisorbed Al geometries [152-158]. None of them proved compatible with the measured ELEED intensities: The ELEED intensity analysis revealed, instead, that for Θ = 0.5 ML Al replaced Ga in the second layer beneath the interface [63,148]: a controversial conclusion which later was confirmed both by energy-minimization calculations [140-143] and by core-level photoemission spectroscopy. The low-coverage analysis subsequentlywas extended to encompass a series of GaAs(110)-p(1 × 1)-Al(Θ) structures for 0 < Θ < 8.5 ML [63]. With increasing coverage, Θ, Al first displaces Ga in the second layer, then the third and finally the first forming a thin epitaxical layer of AlAs(110) on GaAs(110). A similar result occurs for Al on GaP(110) [159].

These results admit a simply interpretation which has provided the basis for a practical technology of metal source and drain contacts on thin film transistors [160]. Upon annealing Al deposited on GaAs(110), Al replaces the Ga first in the second atomic layer, next in the third, then in the first, revealing a complicated interplay between the energetically favored bulk Al-for-Ga replacement reaction and the kinetic limitations on Al and Ga diffusion. Moreover, the AlAs surface layer forms a cap on the GaAs substrate, hindering further diffusion in accordance with known activation energies [161] and measurements on Au/GaAs contacts [162]. Thus, we are led to the concept that the deposition and reaction of a few monolayers of reactive metal onto a semiconductor substrate can lead to a diffusion barrier at the interface and to profound changes in its electrical properties. This concept was utilized in the early 1980's to solve fabrication problems associated with large-area arrays of CdSe field effect transistors [160], thereby providing an illustration of the intimate connection between the science of semiconductor surface reconstruction and technology of semiconductor electronics.

3 Wurtzite Cleavage Surfaces

In contrast to zincblende structure materials for which (110) is the only cleavage surface, wurtzite structure compound semiconductors exhibit two cleavage faces: the $(10\bar{1}0)$ and $(11\bar{2}0)$ surfaces. Many common II-VI compounds exhibit the wurtzite structures, e.g., ZnO, ZnS, CdS and CdSe. Apart from a long historical interest in ZnO as a catalyst [163], a prototypical "ionic" semiconductor [50,51,96,164-166], and a sensor material [167], attention to the surfaces of wurtzite structure compound semiconductors is recent in origin. In particular, they constitute an important test of the generality of the concepts developed in the preceding section for zincblende structure materials, i.e., the universality of the surface atomic geometries and the surface-state lowering mechanism for surface reconstructions of the cleavage faces of II-VI compounds. Initial theoretical studies [16] predicted both phenomena for wurtzite cleavage faces, and recent experiments [30-32,98] are validating these predictions. We consider each of the two cleavage surfaces, in turn, below.

3.1 Wurtzite $(10\bar{1}0)$

The unrelaxed $(10\bar{1}0)$ surfaces of wurtzite structure compound semiconductors exhibit an anion-cation dimer structure with each surface species bonded to two atoms in the layer beneath. Diagrams and descriptions of this surface may be found in early reviews [23,25,168]. Upon cleavage, however, these surfaces relax due to the redistribution of dangling-bond charge via approximately bond-length-conserving top-layer rotations analogous to the zincblende (110) surface [16,21]. A picture of the predicted [21] relaxed surface already has been given in Fig. 5. The independent surface structure parameters and associated reciprocal lattice (i.e., normal incidence LEED pattern)

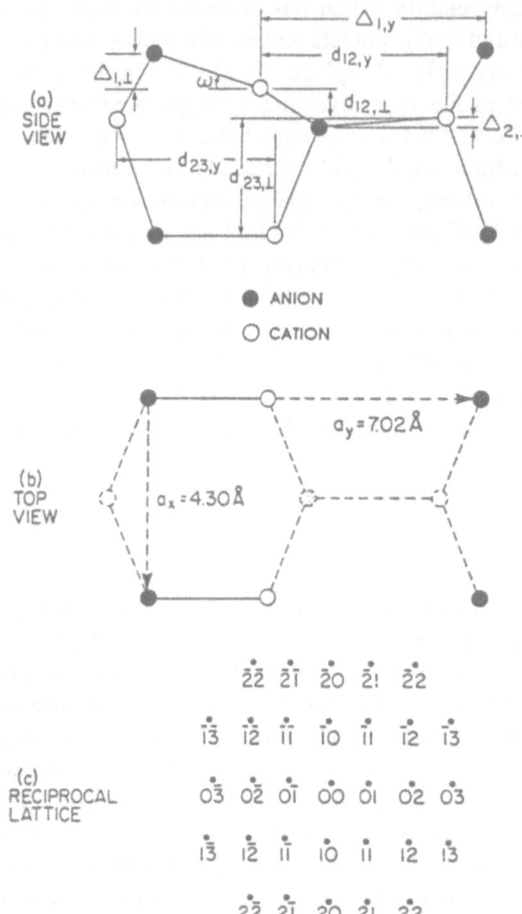

Figure 14 Schematic diagram of the independent structural variables and the associated reciprocal lattice for $(10\bar{1}0)$ surfaces of wurtzite structure compound semiconductors. Panel (a): Side view. Panel (b): Top view. Panel (c): reciprocal lattice. (After Duke et al. [32].)

are indicated in Fig. 14. The qualitative features of the predicted relaxation have been confirmed by an early (1978) study of ZnO$(10\bar{1}0)$ [29, 50, 51] which, however, embodies techniques and data which have been greatly improved in the ensuing decade. Therefore a study of CdSe$(10\bar{1}0)$ was recently undertaken using both LEED [30,31] and Low-energy position diffraction (LEPD) [31,32]. The appropriate dimensionless tilt angle, ω, shown in Fi. 14 is redicted to be 18° for CdSe$(10\bar{1}0)$ [97]. The LEED intensity analysis yields $\omega = 22° \pm 4°$ [30,31] and the LEPD analysis gives $\omega = 15° \pm 5°$ [31,32]. Thus, both analyses give results compatible with the theoretical prediction.

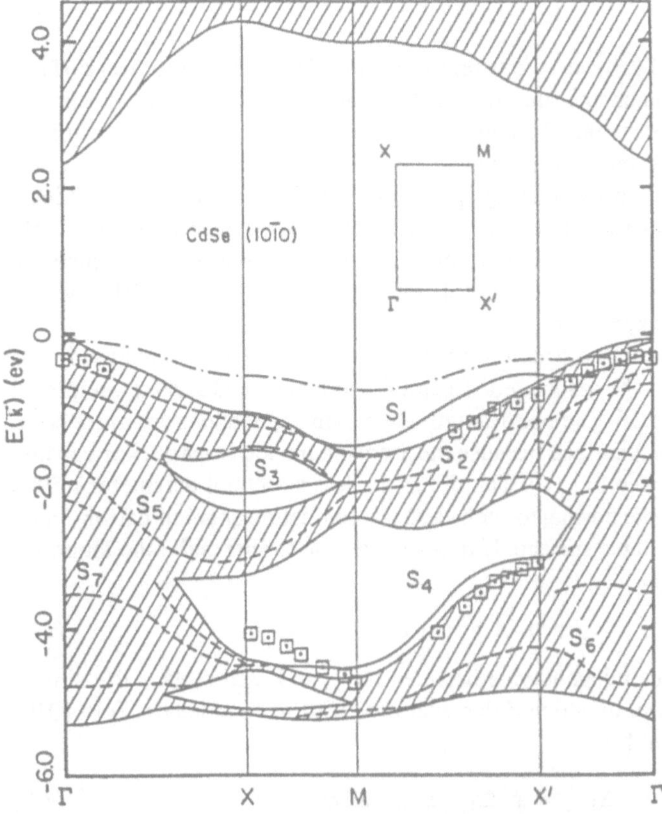

Figure 15 Surface electronic band structure of CdSe(10$\bar{1}$0). The surface Brillouin zone is indicated in the inset. The surface-state eigenvalues are denoted by solid lines and the surface-resonance eigenvalues by the dashed lines. The dot-dashed line designates the energies surface states S1 characteristic of the unreconstructed surface. The squares are the surface-state energies extracted from angle-resolved photoelectron spectra. The states are labeled according to Wang and Duke [97]. (After Wang et al. [98].)

An examination of the surface-state eigenvalue spectrum via ARPES also has been reported [98] with the results shown in Fig. 15. The predictions are in satisfactory correspondence with the surface-state energies determined from the ARPES data, indicating that the surface-state lowering mechanism is appropriate for CdSe(10$\bar{1}$0) as well as for zincblende (110) surfaces. The rotational surface phonon energies have been calculated for the wurtzite $(10\bar{1}0)$ surfaces [28,169] but have not yet been observed experimentally.

3.2 Wurtzite $(11\bar{2}0)$

The unrelaxed wurtzite $(11\bar{2}0)$ surface is reminiscent of the zincblende (110) surface. It consists of anion-cation chains such that each surface species exhibits two surface bonds and one subsurface bond. In addition, however, wurtzite $(11\bar{2}0)$ exhibits a new feature: four (rather than two) atoms per surface unit cell and a glide-plane symmetry which is observed via missing spots in the LEED pattern [23-25]. Indeed, the observations of the missing spots led to early speculation that ZnO $(11\bar{2}0)$ might be unrelaxed [170] although subsequent calculations [96] revealed the occurrence of relaxations, analogous to those of zincblende (110) and wurtzite $(10\bar{1}0)$, which preserved the glide-plane symmetry.

The predicted relaxed surface structure for the wurtzite $(11\bar{2}0)$ surface already has been illustrated in Fig. 6. The associated structural parameters and surface reciprocal lattice are specified in Fig.16. The bond-length-conserving relaxations are more complicated than for the $(10\bar{1}0)$ surface because the atoms in the top layer pucker causing it to become non-planar (see Fig.16). The dimensionless structural parameter in this case is the angle between the normal to each cation-anion-cation or anion-cation-anion triplet and that of the unrelated surface. If n designates the former and z the latter, we have for the polar angle, ω:

$$n \cdot z = \cos\omega \tag{3.1}$$

This leads to a complicated dependence between $\Delta_{1,\perp}$ and ω to replace the simple expression $\Delta_{1,\perp} = d\sin\omega$, in which d designates the Cd-Se bond length, for $(10\bar{1}0)$ surface. It is of the general form

$$\Delta_{1,\perp} = f(\Delta_{1,x}\Delta_{1,y})\tan\omega \tag{3.2}$$

in which variations in $\Delta_{1,\perp}$, $\Delta_{1,x}$, and $\Delta_{1,y}$ are constrained to keep all the bond lengths constant. These constraints can be satisfied, however, for $0 < \omega < 45°$. Predicated values occur in the vicinity of $\omega = 32°$ [32,97]. The LEED analysis for CdSe$(11\bar{2}0)$ yields $\omega = 33° \pm 4°$ [31] and the LEPD analysis gives $\omega = 27° \pm 7°$ [31,32]. Therefore both LEED and LEPD give structures for CdSe$(11\bar{2}0)$ which are in quantitative agreement with the concept that it exhibits a bond-length-conserving top-layer relaxation leading to a distorted p^3 conformation for the anion and sp^2 conformation for the cation. Since no structural data for wurtzite $(11\bar{2}0)$ were available prior to the model predictions [97], the experimental results for CdSe$(11\bar{2}0)$ constitute an *a priori* verification of the surface-state lowering mechanism for the cleavage surfaces of tetrahedrally coordinated compound semiconductors. The predicted [16] near-invariance of ω with varying materials has not yet been tested experimentally.

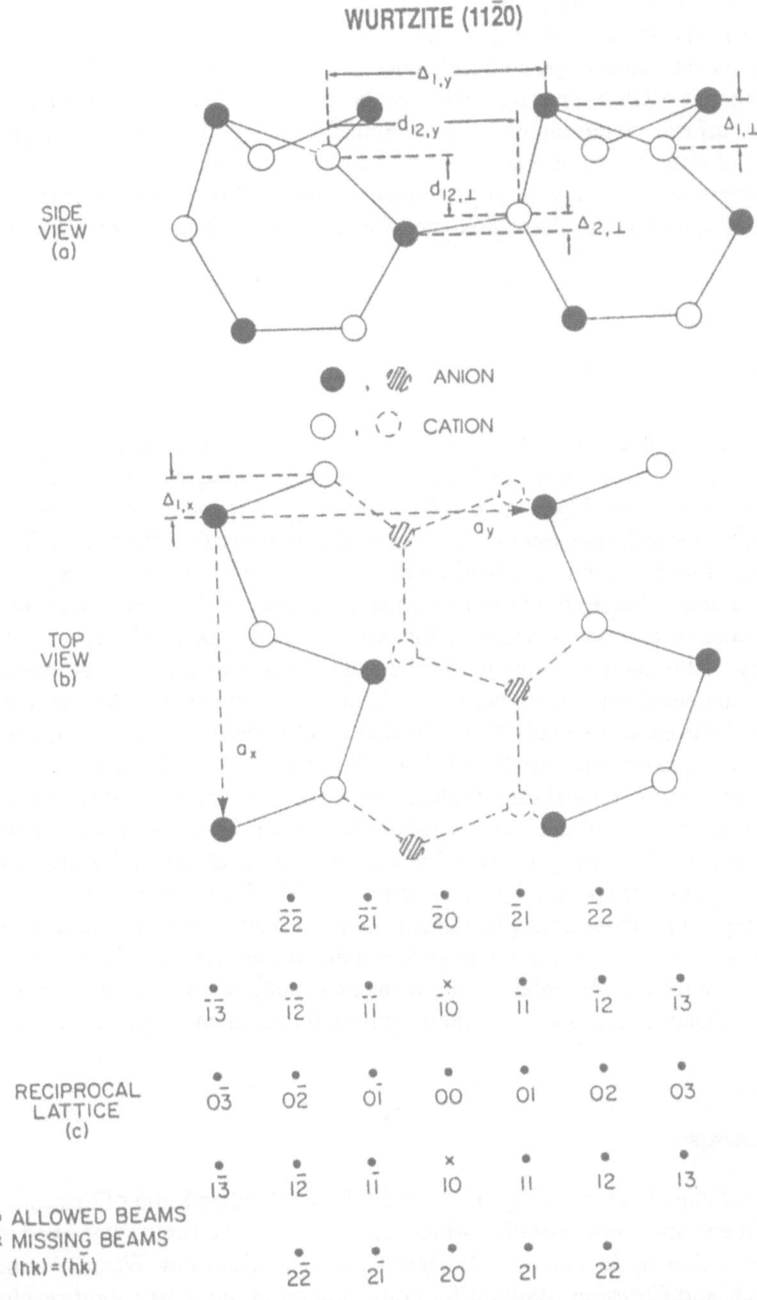

Figure 16 Schematic diagram of the independent structural variables and the associated reciprocal lattice for the (1120) surface of wurtzite structure compound semiconductors. Panel (a): Side view. Panel (b): Top view. Panel (c): reciprocal lattice. (After Duke et al. [32].)

Calculations of surface-state energies for the $(11\bar{2}0)$ surfaces of ZnO [96], ZnS [42], ZnSe [43], CdS [97] and CdSe [97,98] indicate that the surface states characteristic of the truncated bulk surface disappear entirely upon relaxation, becoming surface resonances. Although no ARPES measurements have been reported for $(11\bar{2}0)$ compound semiconductor surfaces, angle-integrated photoemission energy distributions suggest the occurrence of strong surface resonances at the top of the valence band [98]. No surface phonon calculations have been performed for the $(11\bar{2}0)$ surface. Surface rotational phonons analogous to those predicted for the $(10\bar{1}0)$ surfaces are expected [28].

4 Synopsis

The reconstructions of the cleavage faces of tetrahedrally coordinated compound semiconductors are of particular importance in surface science for three reasons. First, they reveal the occurrence of novel types of chemical bonding not found in either molecules or bulk solids. Second, they exhibit approximately "universal" surface structures characterized by dimensionless variables which exhibit common values for a given surface of all materials. These structures reveal the dominance of the effects of atomic chemistry by those of atomic connectivity for this class of surfaces. Third, a single coherent theory of the atomic geometry, surface electronic structure and atomistic dynamics has been developed which permits the quantitative prediction and interpretation of all available experimentally determined results for tetrahedrally-coordinated compound semiconductors: both III-V and II-VI. Therefore these surfaces have been the locus of the discovery of novel surface chemical phenomena and the testing ground for theoretical constructs leading to successful predictions of these phenomena. With the advent of enhanced accuracy of experimental structure analyses and extraction of surface-state eigenvalue spectra from photoemission, EELS and HAS experiments these new concepts will undoubtedly be refined and even revised. Nevertheless, even today, the study of these cleavage surfaces as documented above reveals the power of modern surface science experimental techniques and the methodologies of condensed-matter theory to discover previously unknown physical and chemical phenomena at surfaces.

Acknowledgements

The author would like to express his gratitude to Professor Antoine Kahn of Princeton University for the years of collaboration which led to many of the results reviewed in this manuscript. I also am indebted to Dr. John LaFemina, Yao Rong Wang, Wayne Ford, Del Lessor and Christian Mailhiot for collaboration in the most recent results treated herein and for the use of material which was unpublished at the time of writing the original draft of this review.

Bibliography

[1] C. B. Duke, in Surface Properties of Electronic Materials, ed. by D. A. King and D. P. Woodruff (Elsevier Science Publishers, Amsterdam, 1988), pp. 69-118.

[2] A. Kahn, Surf. Sci. Repts. **3**, 193(1983).

[3] C. B. Duke and Y. R. Wang, J. Vac. Sci. Technol. A **7**, 2035(1989).

[4] C. B. Duke, S. L. Richardson, A. Paton and A. Kahn, Surf. Sci. **127**, L135(1983).

[5] M. W. Puga, G. Xu, and S. V. Tong, Surf. Sci. **164**, L789(1985).

[6] C. B. Duke and A. Paton, Surf. Sci. **164**, L797(1985).

[7] L. Smit, T. E. Derry and J. F. van der Veen, SurfSci. **150**, 245(1985).

[8] H. J. Gossman and W. M. Gibson, Surf. Sci. **139**, 239(1984).

[9] D. R. Hamann, Phys. Rev. Lett. **46**, 1227(1981). 1

[10] R. M. Feenstra and A. P. Fein, Phys. Rev. B **32**, 1394(1985).

[11] R. Blumenthal, S. K. Donner, J. L. Herman, R. Trehan, K. P. Caffey, E. Furman, N. Winograd and B. D. Weaver, J. Vac. Sci. Technol. B **6**, 1444(1988).

[12] C. Mailhiot, C. B. Duke and Y. C. Chang, Phys. Rev B **30**, 1109 (1984).

[13] V. Dose, H.-J. Gossmann and D. Straub, Surf Sci. **117**, 387(1982).

[14] D. Straub, M. Skibowski and F. J. Himpsel, Phys. Rev. B **32**, 5237 (1985).

[15] C. B. Duke, J. Vac Sci. Technol. B **1**, 732(1983).

[16] C. B. Duke and Y. R. Wang, J. Vac. Sci. Technol. A **6**, 692(1988).

[17] C. A. Swarts, W. A. Goddard and T. C. McGill, J. Vac. Sci. Technol. **17**, 982(1980).

[18] C. B. Duke, R. J. Meyer and P. Mark, J. Vac. Sci. Technol. **17**, 971 (1980).

[19] M.-H. Tsai, J. D. Dow, R. P. Wang and R. V. Kasowski, Phys. Rev. B. **40**, 9818(1989).

[20] C. B. Duke, in Atomic and Molecular Processing of Electronic and Ceramic Materials, ed. by I. A. Aksay, G. L. McVay, T. G. Stroebe and J. F. Wagner (Materials Research Society, Pittsburgh, 1987), pp. 3-10.

[21] C. B. Duke and Y. R. Wang, J. Vac. Sci. Technol. B **6**, 1440(1988).

[22] A. U. MacRae and G. B. Gobeli, J. Appl. Phys. **35**, 1629(1964).

[23] P. Mark, S. C. Chang, W. F. Creighton and B. W. Lee, Crit. Rev. Solid. State Sci. **5**, 189(1975).

[24] A. R. Lubinsky, C. B. Duke, B. W. Lee and P. Mark, Phys. Rev. Lett. **36**, 1058(1976).

[25] C. B. Duke, Crit. Rev. Solid State Mater. Sci. **8**, 69(1-978).

[26] P. Mark, A. Kahn, G. Cisneros, and M. Bonn, Crit. Rev. Solid State Matter. Sci. **8**, 317(1979).

[27] C. B. Duke, J. Vac. Sci. Technol. A **6**, 1957(1988).

[28] C. B. Duke and Y. R. Wang, J. Vac. Sci. Technol. B **7**, 1027(1989).

[29] C. B. Duke, R. J. Meyer, A. Paton and P. Mark, Phys. Rev B **18**, 4225 (1978).

[30] C. B. Duke, A. Paton, Y. R. Wang, K. Stiles and A. Kahn, Surf. Sci. **197**, 11(1988).

[31] T. N. Horsky, G. R. Brandes, K. F. Canter, C. B. Duke, S. F. Horng, A. Kahn, D. L. Lessor, A. P. Mills, Jr., A. Paton, K. Stevens and K. Stiles, Phys. Rev. Lett. **62**, 1876(1989).

[32] C. B. Duke, D. L. Lessor, T. N. Horsky, G. Brandes, K. F. Canter, P. H. Lippel, A. P. Mill, Jr., A. Paton and Y. R. Wang, J. Vac. Sci. Technol. A **7**, 2030(1989).

[33] R. Blumenthal, S. K. Donner, J. L. Herman, R. Trehan, K. P. Caffey, E. Furman, N. Winograd, and B. D. Weaver, J. Vac. Sci. Technol. B **6**, 1444(1988).

[34] P. G. Cowell, M. Prutton and S. P. Tear, Surf. Sci. **177**, L915(1986).

[35] V. E. de Carvalho, M. Prutton and S. P. Tear, Surf. Sci. **184**, 198 (1987)

[36] P. G. Cowell and V. E. de Carvalho, Surf. Sci. **187**, 175(1987).

[37] P. G. Cowell and V. E. de Carvalho, J. Phys. C: Solid State Phys. **21**, 2983(1988).

[38] D. J. Smith, R. W. Glaisher and P. Lu, Phil. Mag. Lett. **59**, 69(1989).

[39] U. Harten and P. Toennies, Europhys. Lett. **4**, 833(1987).

[40] R. B. Doak and D. B. Nguyen, J. Electron-Spectros. Rel. Phenom. **44**, 205(1987).

[41] Y. R. Wang and C. B. Duke, Surf. Sci. **205**, L755(1988).

[42] Y. R. Wang and C. B. Duke, Phys. Rev. B **36**, 2763(1987).

[43] Y. R. Wang and C. B. Duke and C. Mailhiot, Surf. Sci. **188**, L708 (1987).

[44] Y. R. Wang, C. B. Duke, K. O. Magnusson and S. A. Flodström, Surf. Sci. **205**, L760(1988).

[45] A. C. Ferraz and G. P. Srivastava, Surf. Sci. **182**, 161(1987).

[46] G-X. Qian, R. M. Martin and D. J. Chadi, Phys. Rev. B **37**, 1303 (1988).

[47] A. C. Ferraz and G. P. Srivastava, J. Phys. C: Solid State Phys. **19**, 5987(1986).

[48] R. V. Kasowski, M.-H. Tsai and J. D. Dow, J. Vac. Sci. Technol. B **5**, 953(1987).

[49] M.-H. Tsai, J. D. Dow, R. P. Wang and R. V. Kasowski, Superlattices and Microstructures **6**, 431(1989).

[50] C. B. Duke, A. R. Lubinsky, B. W. Lee and P. Mark, J. Vac. Sci. Technol. **13**, 761(1976).

[51] C. B. Duke, A. R. Lubinsky, S. C. Chang, B. W. Lee and P. Mark Phys. Rev. B **15**, 4865(1977).

[52] C. B. Duke, A. Paton and A. Kahn, J. Vac. Sci. Technol. A **1**, 672 (1983).

[53] C. B. Duke, A. Paton and A. Kahn, J. Vac. Sci. Technol. A **2**, 515 (1984).

[54] J. J. Joyce, J. Anderson, M. M. Nelson, C. Yu and G. J. Lapeyre, J. Vac. Sci. Technol. A **7**, 850(1989).

[55] A. B. McLean, R. M. Feenstra, A. Taleb-Ibrahimi, and R. Ludeke, Phys. Rev. B **39**, 12925(1989).

[56] Y. Hu, T. J. Wagener M. B. Jost and J. H. Weaver Phys. Rev. B **40**, 1146(1989).

[57] R. Ludeke, A. Taleb-Ibrahimi, R. M. Feenstra and A. B. McLean, J. Vac. Sci. Technol. B **7**, 936(1989).

[58] C. B. Duke, D. L. Lessor, T. Guo and W. K. Ford, J. Vac. Sci. Technol. A **8**, xxx (1990).

[59] C. B. Duke, in Solvay Conference on Surface Science, ed. by F. W. de Wette (Springer Verlag, Berlin, 1988), pp. 361-365.

[60] C. B. Duke, A. Paton, A. Kahn and D.-W. Tu, J. Vac. Sci. Technol. B 2 366(1984).

[61] C. B. Duke, A. Paton, W. K. Ford, A. Kahn and G. Scott, Phys. Rev. B **24**, 3310(1981).

[62] C. B. Duke, A. Paton, A. Kahn and C. R. Bonapace, Phys. Rev. B **28**, 852(1983).

[63] A. Kahn, J. Carelli, D. Kanani, C. B. Duke, A. Paton and L. J. Brillson, J. Vac. Sci. Technol. **19**, 331(1981).

[64] C. B. Duke, A. Paton, W. K. Ford, A. Kahn and J. Carelli, Phys. Rev. B **24**, 562(1981).

[65] B. W. Lee, R. K. Ni, N. Masud, X. R. Wang and M. Rowe, J. Vac. Sci. Technol. **19**, 294(1981).

[66] C. B. Duke, A. Paton and A. Kahn, Phys. Rev. B **27**, 3436(1983).

[67] L. Smit, R. M. Tromp and J. F. van der Veen, Phys. Rev B **29**, 4814 (1984).

[68] L. Smit and J. F. van der Veen, Surf. Sci. **166**, **183**, (1986).

[69] R. J. Meyer, C. B. Duke, A. Paton, J. C. Tsang, J. L. Yeh, A. Kahn and P. Mark, Phys. Rev. B. **22**, 6171(1980).

[70] S. P. Tear, M. R. Walton-Cook, M. Prutton and J. A. Walker, Surf. Sci. **99**, 598(1980).

[71] C. B. Duke, A. Paton, A. Kahn and C. R. Bonapace, Phys. Rev. B **27**, 6189(1983).

[72] G. P. Srivastava, I. Singh, V. Montgomery and R. H. Williams, J. Phys. C: Solid State Phys. **16**, 3627(1983).

[73] R. J. Meyer, C. B. Duke, A. Paton, J. L. Yeh, J. C. Tsang, A. Kahn and P. Mark, Phys. Rev. B **21**, 4740(1980).

[74] C. B. Duke, R. J. Meyer, A. Paton, A. Kahn, J. Carelli and J. L. Yeh, J. Vac. Sci. Technol. **18**, 866(1981).

[75] R. J. Meyer, C. B. Duke, A. Paton, E. 5o, J. L. Yeh, A. Kahn and P. Mark, Phys. Rev. B **22**, 2875(1980).

[76] R. P. Beres, R. E. Allen and J. D. Dow, Phys. Rev. B **26**, 769(1982).

[77] D. J. Miller and D. Haneman, Phys. Rev. B **3**, 2918(1971).

[78] D. J. Miller and D. Haneman, J. Vac. Sci. Technol. **15**, 1267(1978).

[79] R. P. Beres, R. E. Allen, J. P. Buisson, M. A. Bowen, G. F. Blackwell, H. P. Hjalmarson and J. D. Dow, J. Vac. Sci. Technol. **21**, 548(1978).

[80] C. A. Swarts, T. C. McGill and W. A. Goddard III, Surf. Sci. **110**, 400 (1981).

[81] R. Chang and W. A. Goddard III, Surf Sci. **114**, 311(1984).

[82] D. J. Miller and D. Haneman, Surf. Sci. **82**, 102(1979).

[83] C. Mailhiot, C. B. Duke and D. J. Chadi, Surf. Sci. **149**, 366(1985).

[84] D. J. Chadi, Phys. Rev. B **19**, 2074(1979).

[85] D. J. Chadi, Phys. Rev. Lett. **41**, 1062(1978).

[86] C. Mailhiot, C. B. Duke and D. J. Chadi, Phys. Rev. B **31**, 2213(1985).

[87] R. P. Beres, R. E. Allen and J. D. Dow, Solid State Commun. **45**, 13 (1983).

[88] F. Manghi, C. M. Bertoni, C. Calandra and E. Molinari, Phys. Rev. B **24**, 6029(1981).

[89] F. Manghi, E. Molinari, C. M. Bertoni and C. Calandra, J. Phys. C: Solid State Phys. **15**, 1099(1982).

[90] R. P. Beres, R. E. Allen and J. D. Dow, Phys. Rev. B **26**, 5702(1982).

[91] M. Schmeits, A. Mazur and J. Pollmann, Solid State Commun. **40**, 1081(1981).

[92] C. Calandra, M. Manghi and C. M. Bertoni, J. Phys. C: Solid State Phys. **10**, 1911(1977).

[93] J. R. Chelikowsky and M. L. Cohen, Phys. Rev. B **13**, 826(1976).

[94] K. O. Magnusson, N. O. Karlsson, D. Straub, S. A. Flodström, and F. J. Himpsel, Phys. Rev. B **36**, 6566(1987).

[95] P. Vogel, H. P. Hjalmarson and J. D. Dow, J. Phys. Chem. Solids **44**, 365(1983).

[96] Y. R. Wang and C. B. Duke, Surf. Sci. **192**, 309(1987).

[97] Y. R. Wang and C. B. Duke, Phys. Rev. B **37**, 6417(1988).

[98] Y. R. Wang, C. B. Duke, K. Stevens, A. Kahn, K. O. Magnusson and S. A. Flodström, Surf. Sci. **206**, L817(1988).

[99] W. A. Harrison, Electronic Structure and the Properties of Solids (Freeman, San Francisco, 1980).

[100] C. Messmer and J. C. Bilello, J. Appl. Phys. **52**, 4623(1981).

[101] R. J. Meyer, C. B. Duke, A. Paton, A. Kahn, E. So, J. L. Yeh and P. Mark, Phys. Rev. B **19**, 5194(1979).

[102] S. Y. Tong, W. N. Mei and G. Xu, J. Vac. Sci. Technol. B **2**, 393(1984).

[103] A. Ebina, T. Unno, Y. Suda, H. Koinuma and T. Takahashi, J. Vac. Sci. Technol. **19**, 301(1981).

[104] T. Takahashi and A. Ebina, Appl. Surf. Sci. **11/12**, 268(1982).

[105] A. Mazur and J. Pollmann, Phys. Rev. B **39**, 5261(1989).

[106] D. C. Allan and E. J. Mele, Phys. Rev. Lett. **53**, 826(1984).

[107] O. L. Alerhand and E. J. Mele, Phys. Rev. B **35**, 5533(1987).

[108] O. L. Alerhand and E. J. Mele, Phys. Rev. Lett. **59**, 657(1987).

[109] O. L. Alerhand and E. J. Mele, Phys. Rev. B **37**, 2536(1988).

[110] L. Miglio. P. Santini, P. Ruggerone and G. Benedek, Phys. Rev. Lett. **62**, 3070(1989).

[111] J. Carelli and A. Kahn, Surf. Sci. **116**, 380(1982).

[112] C. B. Duke, A. Paton, W. K. Ford, A. Kahn and J. Carelli, Phys. Rev. B **26**, 803 (1982).

[113] K. Li and A. Kahn, J. Vac. Sci. Technol. A **4**, 958 (1986).

[114] W. Pletschen, N. Esser, H. Munder, D. Zahn, J. Geurts and W. Richter, Surf Sci. **178**, 140 (1986).

[115] M. Mattern-Klossen, R. Strümpler and H. Lüth, Phys. Rev. B **33**, 2259 (1986).

[116] A. Tulke, M. Mattern-Klossen and H. Lüth, Solid State Commun. **59**, 303 (1986).

[117] A. Tulke and H. Lüth, Surf. Sci. **178**, 131 (1986).

[118] R. Strümpler and H. Lüth, Surf. Sci. **182**, 545 (1987).

[119] P. Mårtensson, G. V. Hansson, M. Lähdeniemi, K. O. Magnusson, S. Wiklund and J. M. Nicholls, Phys. Rev. B **33**, 7399 (1986).

[120] F. Schäffler, R. Ludeke, A. Taleb-Ibrahimi, G. Hughes and D. Rieger, J. Vac. Sci. Technol. B **5**, 1046 (1987).

[121] F. Schäffler, R. Ludeke, A. Taleb-Ibrahimi, G. Hughes and D. Rieger, J. Vac. Sci. Technol. B **36**, 1328 (1987).

[122] R. M. Feenstra and P. Mårtensson, Phys. Rev. Lett. **61**, 447 (1988).

[123] P. Mårtensson and R. M. Feenstra, Phys. Rev. B **39**, 7744 (1989).

[124] W. Drube and F. J. Himpsel, Phys. Rev. B **37**, 855 (1988).

[125] C. Maani, A. C. McKinley and R. H. Williams, J. Phys. C: Solid State Phys. **18**, 4975 (1985).

[126] C. B. Duke, C. Mailhiot, A. Paton, K. Li, C. Bonapace and A. Kahn, Surf Sci. **163**, 391 (1985).

[127] D. Zahn, N. Esser, W. Pletschen, J. Geurts and W. Richter, Surf Sci. **168**, 823 (1986).

[128] R. H. Williams, D. Zahn, N. Esser and W. Richter, J. Vac. Sci. Technol. B **7**, 997 (1989).

[129] T. Kendelewitz, R. Cao, K. Miyano, I. Lindau and W. E. Spicer, J. Vac. Sci. Technol. A **7**, 765 (1989).

[130] T. Kendelewitz, R. Cao, K. Miyano, I. Lindau and W. E. Spicer, J. Vac. Sci. Technol. B **7**, 991 (1989).

[131] J. J. Joyce, J. Anderson, M. M. Nelson and G. J. Lapeyre, Phys. Rev. B **40**, 10412 (1989).

[132] A. B. McLean and F. J. Himpsel, Phys. Rev. B**40**, 8425 (1989).

[133] W. K. Ford, T. Guo, S. L. Lantz, K. Wan, S.-L. Chang, C. B. Duke and D. L. Lessor, J. Vac. Sci. Technol B **8**, xxxx (1990).

[134] C. Mailhiot, C. B. Duke and D. J. Chadi. Phys. Rev. Lett. **53**, 2114 (1984).

[135] P. Skeath, I. Lindau, C. Y. Su and W. E. Spicer, J. Vac. Sci. Technol. **19**, 556 (1981).

[136] J. P. La Femina, C. B. Duke and C. Mailhiot, J. Vac. Sci. Technol. B **8**, xxxx (1990).

[137] C. M. Bertoni, C. Calendra, F. Manghi and E. Molinari, Phys. Rev. B **27**, 1251 (1983).

[138] A. Zunger, Thin Solid Films **104**, 301 (1983).

[139] P. Skeath, I. Lindau, C. Y. Su and W. E. Spicer, Phys. Rev. B **28**, 7051 (1983).

[140] C. R. Bonapace, K. Li and A. Kahn, J. de Physique 45 **C5**, 409 (1984).

[141] J. Ihm and J. D. Joannopoulos, Phys. Rev. Lett. **47**, 679 (1981).

[142] J. Ihm and J. D. Joannopoulos, J. Vac. Sci. Technol. **21**, 340 (1982).

[143] J. Ihm and J. D. Joannopoulos, Phys. Rev. B **26**, 4429 (1982).

[144] R. R. Daniels, A. D. Katnani, T.-X. Zhao, G. Margaritondo and A. Zunger, Phys. Rev. Lett. **49**, 895 (1982).

[145] A. Zunger, Phys. Rev. B. **24**, 4372 (1981).

[146] A. Zunger, J. Vac. Sci. Technol. **19**, 690 (1981).

[147] P. Skeath, I. Lindau, C. Y. Su, P. W. Chye and W. E. Spicer, J. Vac. Sci. Technol. **17**, 511 (1980).

[148] C. B. Duke, A. Paton, R. J. Meyer, L. J. Brillson, A. Kahn, D. Kanani, J. Carelli, J. L. Yeh, G. Margaritondo and A. D. Katani, Phys. Rev. Lett. **46**, 440 (1981).

[149] A. Kahn, D. Kanani, J. Carelli, J. L. Yeh, C. B. Duke, R. J. Meyer, A. Paton and L. J. B ril Ison, J. Vac. Sci. Technol. **18**, 792 (1981).

[150] R. Kern, G. Le Lay and J. J. Metois in Current Topics in Materials Science, ed. by E. Kaldis (North Holland, New York, 1979), chap. 3.

[151] R. Ludeke, J. Vac. Sci. Technol. B **2**, 400 (1984).

[152] J. R. Chelikowsky, S. G. Lovie and M. L. Cohen, Solid State Commun. **20**, 641 (1976).

[153] L. J. Brillson, R. Z. Bachrach, R. S. Bauer and J. McMenamin, Phys. Rev. Lett. **42**, 397 (1979).

[154] D. J. Chadi and R. Z. Bachrach, J. Vac. Sci. Technol. **16**, 1159 (1979).

[155] E. J. Mele and J. D. Joannopoulos, Phys. Rev. Lett. **42**, 1094 (1979).

[156] E. J. Mele and J. D. Joannopoulos, J. Vac. Sci. Technol. **16**, 1154 (1979).

[157] J. J. Barton, C. A. Swarts, W. A. Goddard III and T. C. McGill, l. Vac. Sci. Technol. **17**, 164 (1980).

[158] C. A. Swarts, J. J. Barton, W. A. Goddard III and T. C. McGill, J. Vac. Sci. Technol. **17**, 869 (1980).

[159] A. Kohn, C. R. Bonapace, C. B. Duke and A. Paton, J. Vac. Sci. Technol. B **1**, 613 (1983).

[160] C. B. Duke, Appl. Surf. Sci. 11/**12**, 1 (1982).

[161] R. M. Flemming, D. B. McWhan, A. C. Gossard, W. Wiegmann and R. A. Logan, J. Appl. Phys. **51**, 357 (1980).

[162] L. J. Brillson, G. Margaritondo and N. G. Stoffel, Phys. Rev. Lett. **44**, 667 (1980).

[163] W Göpel, Ber. Bunsenges Phys. Chem. **82**, 744 (1978).

[164] J. D. Levine, A. Willis, W. R. Bottoms and P. Mark, Surf. Sci. **29**, 144 (1972).

[165] I. Ivanov and J. Pollmann, Phys. Rev. B **24**, 7275 (1981).

[166] W. Göpel, J. Pollmann, I. Ivanov and B. Reihl, Phys. Rev. B **26**, 3144 (1982).

[167] W. Göpel, Prog. Surf. Sci. **20**, 9 (1985).

[168] C. B. Duke, J. Vac. Sci. Technol. **14**, 870 (1977).

[169] Y. R. Wang and C. B. Duke, Phys. Rev. B **39**, 5569 (1989).

[170] A. R. Lubinsky, C. B. Duke, S. C. Chang, B. W. Lee and P. Mark, J. Vac. Sci. Technol. **13**, 189 (1976).

Resonant Tunneling Diodes: The Effect of Structural Properties on their Performance

Arno Förster

Institut für Schicht- und Ionentechnik
Forschungszentrum Jülich, 52425 Jülich, Germany

1 Introduction

One of the most exciting quantum mechanical phenomena in semiconductor physics is the tunneling process of electrons through potential barriers. As is shown in a variety of publications electrons can pass a potential barrier with a finite transmission probability independent of temperature. One can understand this behavior on the basis of quantum mechanics in which the wave character of the electron is important.

One of the most popular quantum mechanical tunnel devices is a double barrier resonant tunneling diode (DBRTD). In 1974 the first observation of negative differential resistance (NDR) in heterostructures (GaAs/AlGaAs) was found by Chang, Esaki and Tsu [1]. The occurrence of negative differential resistance promised a lot of potential applications and therefore resonant tunneling structures became a subject of current experimental and theoretical interest. Many aspects are the subjekt of controversy such as sequential and resonant tunneling, or the influence of scattering effects especially in indirect band gap material, which are not very clear so far.

The present review briefly discusses some of these points as well as some technological aspects, e. g. growth temperature and interface quality, which severely affect the performance of resonant tunneling diodes (RTDs). The relevance of DBRTD applications has meanwhile been demonstrated by a number of different devices like mixers, oscillators and several high frequency devices. The basis for application is functionality at room temperature.

Although in 1974 the first NDR was observed in heterostructures on the basis of AlGaAs double barriers embedded in GaAs, it took nearly 10 years before the first negative differential resistance at room temperature was measured by Mashahiro Tsuchiya and co-workers [2]. This progress in performance is probably due to the improvement of layer and interface quality in molecular beam epitaxy (MBE). Therefore it seems to be very important to know more details about the influence of growth parameters on the characteristics of the diodes. Especially the role of the interface quality and the scattering mechanisms does not seem to be very clear at the moment. These points will be discussed in sec. 2.3 and sec. 4.

2 General Aspects of Resonant Tunneling

2.1 Tunneling probability

As a result of the wave character of particles, quantum mechanical phenomena become relevant in structures whose dimensions are in the order of nm. A device in which quantum mechanical phenomena are relevant is a potential well. In GaAs wells embedded in AlGaAs electrons can occupy only discrete energy levels whose energetic positions are well understood in terms of quantization effects. The effect is even visible at room temperature and can be used for a variety of practical applications like laser structures. By varying the well width the energy of the quantized states shifts and hence a variation of the photon energy of the emitted light is possible.

Similar quantization effects occur for very thin AlGaAs or AlAs double barriers embedded in GaAs. The corresponding lower conduction band edge as a function of position is shown schematically in Fig. 1.

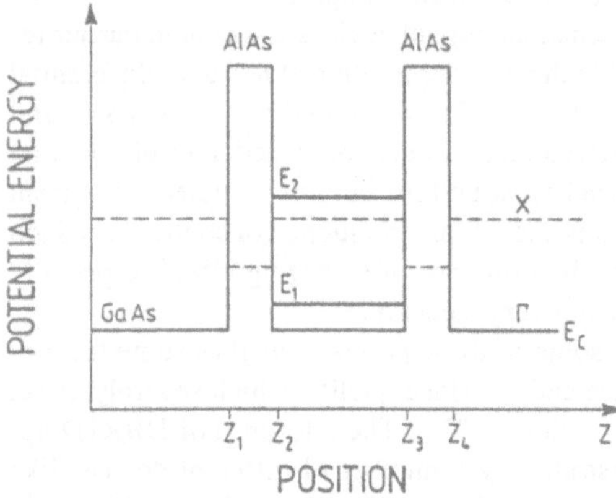

Figure 1
Potential energy profile (conduction band edge) of a symmetrical AlAs double barrier resonant tunneling diode for Γ-electrons (solid line) and X-electrons (dashed line), respectively. E_1 and E_2 denote first and second quasi-bound states in the well.

This double barrier arrangement with barrier thicknesses of only a few monolayers (MLs) is the basis for a resonant tunneling diode. The wave function of the quantized state in the well therefore leaks out through the barriers and couples to moving plane wave states. As a result these states in the well are metastable and the lifetime of the electrons in the well is finite. The electron states are therefore called "quasi-bound states". The lifetime τ of electrons occupying the quasi-bound state is dependent on the half width Γ_0 of the resonance in the well and can be given by the uncertainty principle:

$$\tau = \frac{\hbar}{\Gamma_0}. \tag{1}$$

The transmission probability for electrons is of major interest for such diodes. Electrons having the same energy as the quasi-bound state can pass (i. e. tunnel) through symmetric double barriers with the transmission probability of 1.0, although the tunnel probability for a single barrier is very low. The effect that these electrons do not "see" the barriers is known as the mechanism of resonant tunneling.

Quantum mechanical calculations of the transmission probability are based on a wave function that is coherent over the complete structure. This imposes strict requirements on the crystalline quality of the double barrier resonant tunneling diode. Electrons are not allowed to scatter while passing the double barrier regime. Because of the perfection of interfaces grown by molecular beam epitaxy (MBE) with transition widths of one to two monolayers between GaAs and AlAs and because of the low density of scattering centers in the material it is possible to grow structures in which coherent tunneling processes can be observed even at room temperature.

To look more quantitatively at the coherent tunneling processes through double barriers, calculations of the transmission probability are necessary. Calculations are based on a solution of the Schrödinger Equation:

$$\left(-\frac{\hbar^2}{2}\frac{d}{dz}\frac{1}{m^*(z)}\frac{d}{dz} + \Phi(z)\right)\Psi(z) = E_z\Psi(z) \tag{2}$$

with the electron wave function $\Psi(z)$, the electron energy in z direction E_z, the electron effective mass m^* and the potential energy at the conduction band minimum Φ (z). Suppose a sequence of n different layers with different potential energies Φ_i and different effective masses m_i^* in each layer i. The wave function Ψ_i (z) can be written as a superposition of propagating waves in z and -z direction, with amplitudes A_i and B_i respectively.

Suppose $\Psi_i(z)$ satisfies Eq. (2)

$$\Psi_i(z) = A_i e^{ik_i z} + B_i e^{-ik_i z}. \tag{3}$$

The boundary conditions

$$\Psi_i(z_i) = \Psi_{i+1}(z_i) \tag{4}$$

and the continuity for the wave function

$$\frac{1}{m_i^*}\frac{d}{dz}\Psi_i(z_i) = \frac{1}{m_{i+1}^*}\frac{d}{dz}\Psi_{i+1}(z_i) \tag{5}$$

then leads to a matrix form for the relation between the single plane wave amplitudes:

$$\begin{pmatrix} A_n \\ B_n \end{pmatrix} = \underline{\underline{T}} \begin{pmatrix} A_1 \\ B_1 \end{pmatrix}. \tag{6}$$

The transfer matrix $\underline{\underline{T}}$ relates the amplitudes of the transmitted waves and the incident waves. Since there is only a transmitted wave ($B_n = 0$) the coherent transmission coefficient (T_c) can be written as a ratio of transmitted current to incident current:

$$T_c = \frac{k_n}{k_1} \frac{m_n^*}{m_1^*} \frac{|A_n|^2}{|A_1|^2}. \tag{7}$$

As a result the transmission coefficient T_c follows as:

$$T_c = \frac{k_1}{k_n} \frac{m_n^*}{m_1^*} \frac{1}{|T_{22}|^2} \tag{8}$$

where k_1 and k_n are the wavevectors of the incoming and outgoing waves to and from the double barrier and T_{22} the transfer matrix element [2,2], m_n^* and m_1^* the effective masses on the right and the left side of the double barrier, respectively. This method allows the calculation of the coherent transmission coefficient T_c for a resonant tunneling diode. The conduction band discontinuity between GaAs and AlAs defines the potential energy of the electrons in the barriers. The effective masses are assumed to be constant at each layer.

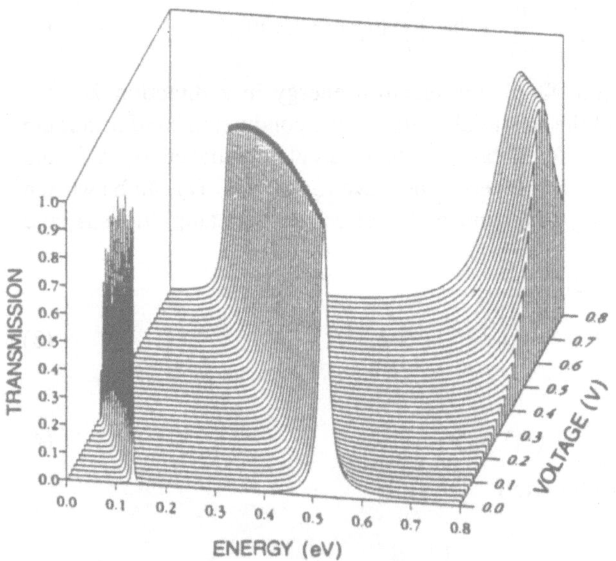

Figure 2
Transmission probability as a function of energy and supply voltage for an AlAs double barrier diode. AlAs barrier thickness is 1.7nm (6 monolayers), GaAs well width is 5 nm. (After Lange [54])

As is shown in Fig. 2, the transmission coefficient directly yields the energy and the half width of the quasi-bound state. Quantitative values for the eigenstates as a function of barrier thickness are given by E. R. Brown [3]. Detailed analyses of resonant tunneling is also given i. e. by B. Ricco [4], M. Ya Azbel [5], Hu and Stapleton [6], Toombs and Sheard [7], and Mendez [8]. The resonance (or quasi bound state) in the coherent transmission probability Tc found at the resonance energy E_r has the form of a Lorentzian [3] with half width Γ_0 as:

$$T_c = T_{res} \frac{\frac{1}{4}\Gamma_0^2}{(E - E_r)^2 + \frac{1}{4}\Gamma_0^2}. \tag{9}$$

The peak transmission is dependent on the symmetry of the double barrier and can be written as (see P. J. Price [11]):

$$T_{res} = \frac{4\Gamma_l \Gamma_r}{(\Gamma_l + \Gamma_r)^2} \tag{10}$$

with partial width Γ_l and Γ_r associated with the transmission properties of the left and the right barriers, respectively. The half width of the resonance is the sum of the corresponding partial widths: $\Gamma_0 = \Gamma_l + \Gamma_r$. In the case of a symmetric double barrier the transmission probability near resonance is 1. According to Eq. (10) the peak transmission for asymmetric double barrier structures is reduced.

It is directly apparent that an applied voltage destroys the symmetry of the barriers and as a result the transmission probability is no longer 1.0. This is shown in Fig. 2 where the simulated Tc is shown as function of the electron energy and the supply voltage. The supply voltage leads to a shift of the resonant states towards the Fermi level which in common structures is several meV above the conduction band edge of GaAs. This resonance can be considered as a channel which opens an electron flux while touching the Fermi level. Therefore the current density increases up to that point. For higher supply voltages the first quasi-bound state is below the conduction band edge of the emitter and no tunneling can occur. This effect causes a negative differential resistance in the I-V characteristics. The current density is strongly affected by the resonance half width Γ_0. This half width (HW) of the Lorentzian is a function of barrier thickness since it is directly related to the degree by which the wave function "leaks out" of the well.

The width of the resonance is exponentially dependent on barrier thickness:

$$\Gamma_0 = c_1 e^{-2\alpha b} \tag{11}$$

with $\alpha = \sqrt{2m_b^*(\Phi_b - E_r)}\,/\hbar$. The function c_1 is dependent on the effective electron mass in the well m_w^* and in the barrier m_b^* with the well width L_w, the potential barrier Φ_b and the resonance energy of the quasi-bound state E_r, whereas α is only dependent on the effective mass in the barrier and the difference between the conduction band offset Φ_b and the resonance energy of the quasi bound state.

In the "thick barrier limit" the function c_1 can be written as (see Brown [3]):

$$c_1 = \frac{16\hbar k \gamma^2}{m_w^*(1+\gamma^2)[(1+\gamma^2)L_w + (1+\gamma^2 m_b^*/m_w^*)2\alpha^{-1}]} \tag{12}$$

with $k = \sqrt{2m_w^* E_r}/\hbar$ and $\gamma = (m_w^*\alpha)/(m_b^* k)$.

2.2 Calculation of current-voltage characteristics

Essential characteristics of resonant tunneling can be derived from the transmission probability as described above. To compare calculations with experimental data simulations of current voltage (I-V) curves are necessary. This first requires a calculation of the transmission probability as a function of electron energy and supply voltage. The supply voltage is assumed to drop linearly in the double barrier regime. Calculations can be done by approximating the actual linear profile into a step-like function of discrete potential values. It was shown by Mendez [8] that this method converges quite rapidly to the exact result. The highly doped supply layers left and right of the double barrier can be described as free electron gases with corresponding Fermi-Dirac distributions.

The current density is calculated as a difference of current densities passing from the left to the right side of the double barrier and vice versa; it can be written after Tsu et al. [9] as:

$$j = \frac{2|e|}{(2\pi)^3} \int_0^\infty dk_z \int_0^\infty dk_{\|}(f_l(E) - f_r(E+eV))T_c(E_z, V)\frac{1}{\hbar}\frac{\partial E_z}{\partial k_z} \tag{13}$$

with the coherent transmission probability $T_c(E_z, V)$ as a function of supply voltage V and energy E_z in z direction and the Fermi distributions f_l and f_r left and right of the double barrier, $k_{\|}$ and k_z denote the parallel- and z-component of the momentum, respectively.

The integration of Eq. (13) leads to the current density with the supply function:

$$j(V) = \frac{4\pi|e|m^* k_b T}{h^3} \int_0^\infty dE_z T_c(E_z, V) \ln \left| \frac{1 + \exp\left(\frac{E_f - E_z}{k_b T}\right)}{1 + \exp\left(\frac{E_f - E_z - eV}{k_b T}\right)} \right| . \tag{14}$$

Calculations of current voltage curves according to Eq. (14) are useful to get a deeper insight in the physics of resonant tunneling diodes. While the peak-current densities resulting from simulations are realistic as compared to experimental values, the calculated valley currents are one or more orders of magnitude lower than the experimental ones. For AlAs/GaAs or AlAs/InGaAs diode structures on GaAs the experimental PVRs at room temperature are in the order of 6. The predicted PVR values from simulations are more than one order of magnitude higher. The reason for this discrepancy is the neglect of scattering in the calculation.

Although there is no unique way of describing scattering in the literature up to now, there is one formally quite simple approach. Scattering is assumed to broaden the resonance in the transmission coefficient while simultaneously damping it. Hence the valley current and the PVR are very strongly influenced. The effect of scattering and especially its influence on the PVR is discussed below.

2.3 Inelastic scattering and half width of resonances

Experimental results by a number of authors have shown that the peak-to-valley ratios (PVRs) achieved in resonant tunneling diodes are strongly dependent on the quality of the interfaces, the substrate temperature during growth, applied hydrostatic pressure and so on. These effects have been attributed, at least partially, to scattering. Inelastic scattering events within the double barrier region lead to a sequential transmission of electrons through the barriers and hence to an incoherent tunneling current. The subject of sequential and resonant tunneling is discussed from different point of views in the literature. Weil and Vinter [10] have postulated, that in the absence of scattering in the well, resonant and sequential tunneling lead to the same DC-current through the structure. Luryi [12] has shown that a negative differential resistance can arise solely from tunneling into a quantum well without any coherent tunneling effect. A corresponding observation was indeed made by Morcoc et al. [13], who reported negative differential resistance by tunneling through a 2.5 nm thick barrier of AlAs into a 5 nm thick GaAs quantum well. The current density was also in the same order of magnitude as that obtained in double barrier structures.

Nevertheless the most appropriate model of scattering is based on the Breit-Wigner generalization of the Lorentzian form of resonant transmission. Büttiker [14] has shown that one can clearly distinguish between resonant and coherent tunneling. Within this formalism resonant tunneling in one dimension is also studied by Stone et al. [15] who derived the total transmission probability in the presence of inelastic scattering for a symmetric structure as:

$$T_{tot} = \frac{\frac{1}{4}\Gamma_0\Gamma}{(E-E_r)^2 + \frac{1}{4}\Gamma^2} \tag{15}$$

with the half width Γ_0 of the eigenstate in the coherent transmission probability T_c and the total resonance half width $\Gamma = \Gamma_0 + \Gamma_i$, where Γ_i represents the inelastic scattering probability. Büttiker [14] interpreted this total transmission probability as a sum of coherent and sequential transmission probability

$$T_{tot} = T_c + T_i. \tag{16}$$

In this picture of scattering the fraction of carriers penetrating the structure coherently is $T_c/T_{tot} = \Gamma_0/\Gamma$ and the fraction of carriers traversing the structure sequentially is $T_i/T_{tot} = \Gamma_i/\Gamma$. From these results one infers that the smaller the elastic width Γ_0, the smaller is the amount of scattering needed to make the sequential tunneling current dominant. This means that in DBRTDs with thick barriers (sharp resonances) in spite of a small scattering probability, considerable sequential tunneling contributions are observed. Furthermore, Eq. (15) can be interpreted as a folding of the coherent transmission probability (Eq. 9) with a normalized Lorentzian of half width Γ_i. The peak transmission is reduced in the presence of inelastic scattering by the factor Γ_0/Γ.

The effect of scattering in this description is shown in Fig. 3, where the first resonance of a resonant tunneling diode is shown for different values of Γ_i (or different inelastic scattering probabilities).

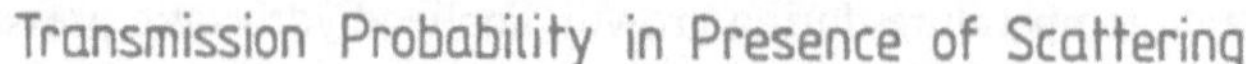

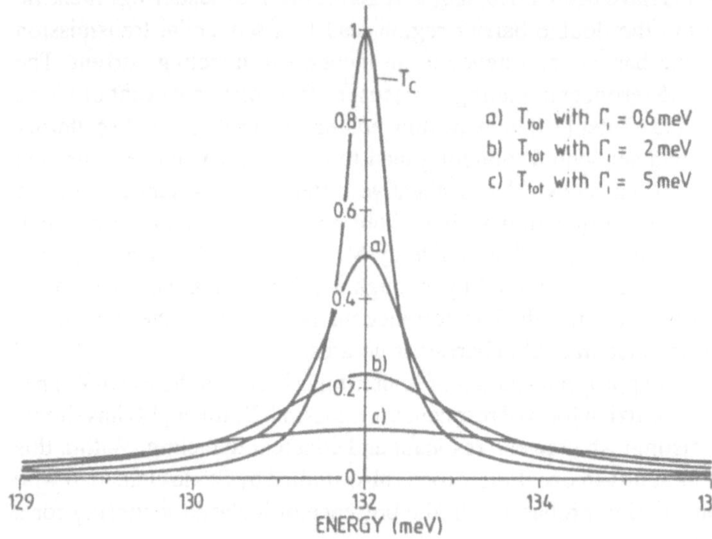

Figure 3
First resonance peak in the transmission probability of Fig. 2 folded with different normalized Lorentzians of half-width Γ_i

From these results one can see that scattering broadens and damps the resonance. This mechanism conserves the peak current density but affects the valley current very strongly resulting in lower PVR values.

Although the scattering mechanism is often not very clear, realistic I-V curves can be calculated according to Eq. (14) with a transmission probability which is folded with a normalized Lorentzian (Fig. 4).

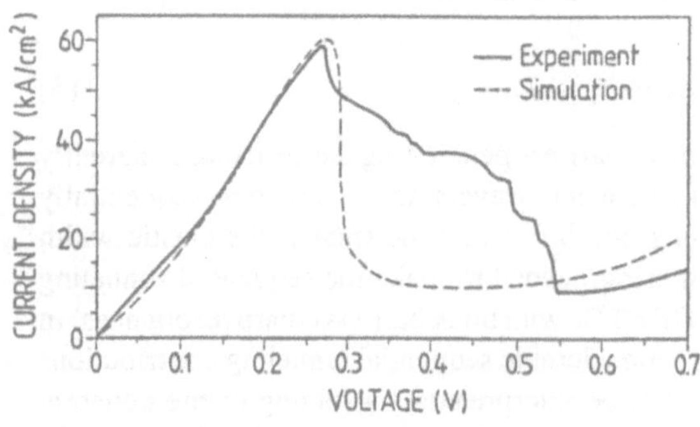

Figure 4
Simulated (solid line) and experimental (dashed line) current voltage characteristics of an AlAs double barrier resonant tunneling diode embedded in $In_{0.17}Ga_{0.83}As$. The simulation was done by folding the transmission probability with a normalized Lorentzian of half width 8 meV. (A series resistance of $8 \times 10^{-7} \Omega/cm^2$ was taken into account). Barrier thickness 6 ML, well width 5 nm. (After Lange [54])

I-V curve simulations lead to the PVR as a function of inelastic scattering parameter

Γ_i (half width of the folded Lorentzian indicating the inelastic scattering probability, see Förster et al. [16]) as:

$$PVR = PVR_0 \frac{\Gamma_0}{\Gamma} \quad (17)$$

with PVR_0 as the PVR without scattering, Γ_0 the half width of the quasi bound state of the transmission probability and $\Gamma = \Gamma_0 + \Gamma_i$ the total half width of the resonance. The PVR in the presence of scattering or sequential tunneling is also suppressed by the factor Γ_0/Γ like the peak transmission in the transmission coefficient by Stone et al [15]. As is shown in Fig. 5, a very small amount of "scattering" (small value of Γ_i) leads to a large reduction in PVR.

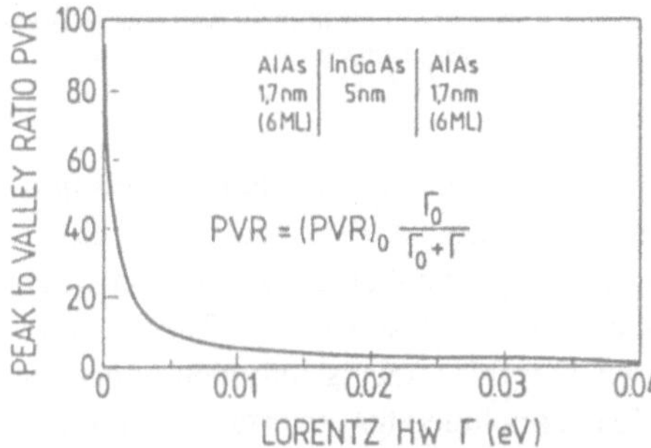

Figure 5
Simulated current peak-to valley (PVR) ratio for the double barrier resonant tunneling diode from Fig. 4 as a function of inelastic scattering probability. The inelastic scattering mechanism is described in the Breit-Wigner formalism by folding the transmission probability with a normalized Lorentzian of half-width Γ_i. (After Förster et al. [16])

Although PVR suppression in the presence of scattering is directly evident from Eq. (17), one cannot learn much about the physical nature of the scattering mechanism itself.

Tyan [17] et al. have used the Breit-Wigner formalism of transmission probability to calculate the effect of longitudinal optical phonon scattering on the tunneling probability. It was shown that phonon scattering very strongly affects the resonance peak. Quantitative results show that the integrated total transmission probability through the potential barriers is not changed, while the transmission peak is reduced and the line width of the resonance is broadened.

Goldman et al. [18] observed a longitudinal optical (LO) phonon satellite peak in the I-V curve of DBRTDs which was simulated by means of LO phonon emission modes by Chevoir and Vinter [19, 20]. This observation is a direct proof of scattering processes in diodes where sequential tunneling is dominant.

A detailed study of scattering effects is given by Chevoir and Vinter [19] who took scattering into account by calculations based on perturbation theory using Fermi's golden rule in the Born approximation. Scattering rates are calculated including interaction with acoustic and optical phonons, alloy disorder scattering in the barrier, interface roughness and impurity scattering. Furthermore the effect of scattering on the current voltage characteristics was calculated for a structure studied by Goldman, Tsui, and Cunningham [18]. The calculations shown in Fig. 6 are based on 5.6 nm GaAs well, 8.5-nm $Al_{0.4}Ga_{0.6}As$ barriers and a doping concentration of $2 \times 10^{17} cm^{-3}$cm^{-3} at 4K.

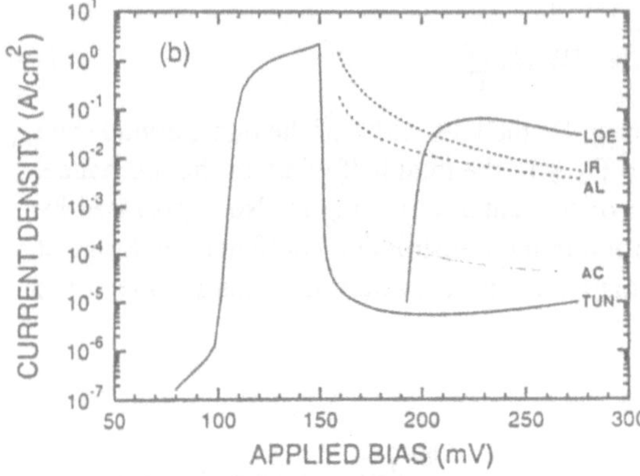

Figure 6
Calculated current-voltage characteristic for an RTD. Following scattering effects are taken into account: optical-phonon emission (LOE) and absorption (LOA), interface roughness (IR), alloy disorder (AL), and acoustic phonons (AC), as well as the direct tunneling transmission (TUN). (After Chevoir and Vinter [19])

It is apparent from the diagram that interface scattering and optical phonon emission are relevant scattering effects in resonant tunneling diodes.

Since the effect of interface roughness is strongly dependent on the interface quality (lateral island size, monolayer thickness of terraces) and the design of the RTD structure itself in a non-trivial way, it is hard to estimate whether this effect is dominant or not. In an AlGaAs double barrier diode Guéret et al. [21] concluded that interface roughness is the dominat scattering mechanism which decreases the PVR.

Furthermore the effect of interface scattering is elucidated in more detail by Leo and McDonald [22] who found significant differences in the effect of interface roughness depending on the location of the rough interface. They concluded that asymmetry effects could also be affected by the interface quality of the diodes.

One of the most discussed scattering mechanisms in resonant tunneling diodes is Γ to X valley scattering. For resonant tunneling diodes built up by combinations of direct and indirect band gap material, the question arises which potential diagram is relevant for the tunneling process. While in the direct band gap material free electrons occupy states near the Γ-point of the band structure, in the indirect material, like AlAs, the lower conduction band edge is located at the X-point in the Brillouin zone. This situation is shown in Fig. 1. The potential diagram for Γ-electrons is depicted by the solid line and for X-electrons by the dotted line, respectively. Hence AlAs acts as a potential barrier for Γ-electrons and as a potential well for X-electrons. From these results it is not clear what potential diagram the electrons really "see" when they are transmitted through the double barrier structure sequentially.

Γ-electrons might tunnel through X-states in AlAs, but scattering processes are required, because Γ- and X- states correspond to different points in the Brillouin zone at different energies and different wavevectors. Therefore this process should be less likely than a $\Gamma - \Gamma$ tunneling process. Indication of resonant tunneling through confined states in GaAlAs at the X-point of the Brillouin zone were found by Mendez et al. [23, 24]. These authors also demonstrated that pressure-dependent measurements of I-V curves are useful to study the Γ-X scattering mechanism [25]. Since hydrostatic pressure affects the band structure differently for the Γ- and the X- gap, there is a pressure-dependent split between Γ- and X- gap energy. According to a review article by Adachi [26] one can estimate the pressure-dependent energy split as E_X = -11 meV / kbar with respect to E_Γ. Indeed Mendez has shown that the performance of the diode is strongly affected by applying hydrostatic pressure. The off-resonance current in $Al_{0.4}Ga_{0.6}As$ double barrier diodes is shown to increase with increasing pressure p thus reducing the PVR.

The effect of hydrostatic pressure on the performance of RTDs is also demonstrated by Brugger et al. [27] (Fig. 7).

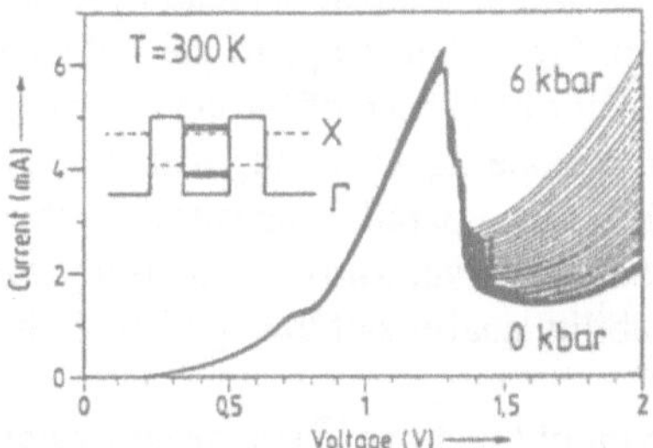

Figure 7
Room temperature current/voltage curves of $Al_{0.7}Ga_{0.3}As/GaAs/Al_{0.7}Ga_{0.3}As$ resonant tunneling diode under hydrostatic pressure. Inset: band scheme for X and Γ conduction band valley. (After Brugger et al. [27])

The current density in the valley region is strongly dependent on the applied hydrostatic pressure. The slope of the curve shows nearly the same pressure dependence that one would expect for a current through X-states on the basis of the Γ-X- split. Current voltage measurements in AlAs double barrier diodes under hydrostatic pressure show that the thinner the barrier the more Γ- tunneling dominates over X- tunneling. The effect of hydrostatic pressure on the performance of resonant tunneling diodes was also studied by Diniz et al. [28] and Brugger et al. [29] for different barrier thicknesses. The authors found a nearly linear dependence of the difference in $j_p - j_v$ as a function of pressure, as is shown in Fig. 8. The high pressure sensitivity of this devices might make RTDs interesting for pressure sensors in high pressure applications.

3 The Stability of the I-V Curve

As is shown in a number of papers, even in DC measurements RTDs can oscillate at very high frequencies and the negative differential resistance (NDR) regime cannot be measured correctly. Often a typical "plateau-like" structure in the NDR region is observed as a result of oscillations. The recording speed is slow compared to the oscillation frequency and therefore only average currents and voltages are monitored. This effect is also discussed by Young et al. [30].

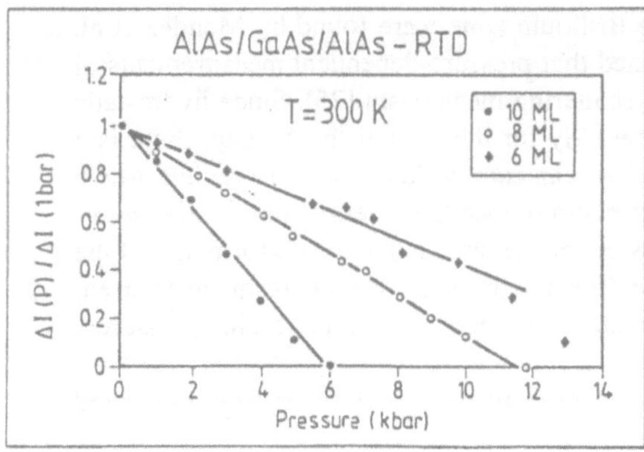

Figure 8
Current swing in the negative-differential resistance region ($\Delta I = I_p - I_v$) as a function of pressure for an AlAs double barrier diode embedded in GaAs for different barrier thicknesses. (After Brugger et al. [29])

A typical I-V curve measured on a diode where oscillations lead to poorly reproducible NDR regions in the plot is shown in Fig. 4. The slope of the I-V curve in the NDR regime is strongly dependent on the recording speed of the curve and especially on the inductance of the supply lines and the diode capacity. Huges et al. [31] showed that in some cases the circuit can be stabilized by a parallel capacitor. Furthermore in some diodes bistabilities are observed in the I-V curve, whose origin is not clear in all cases. It is therefore useful to perform a stability analysis of the diode including series resistance R_s, inductance L and capacity C.

In order to analyze the oscillation behavior of the diode, a simple equivalent circuit is useful to study the characteristics of the diode.

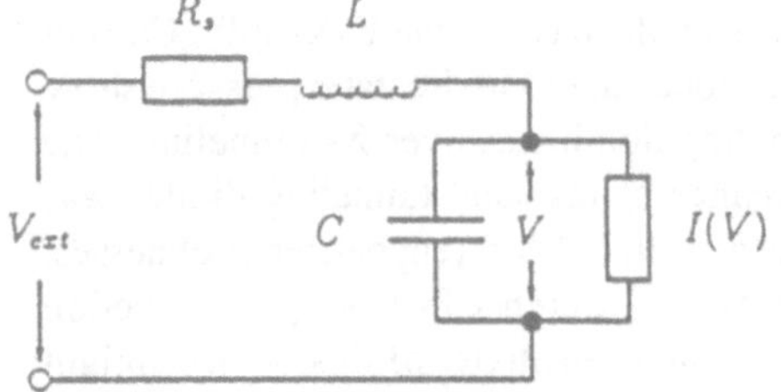

Figure 9
Simple equivalent circuit for a resonant tunneling diode. I(V) characterizes the DC current voltage characteristics of the resonant tunneling diode. R_s, L and C denote the series resistance, the inductance of the supply line and the diode capacity, respectively. V is the intrinsic diode voltage, V_{ext} is the external supply voltage.

An equivalent circuit describing the main features is shown in Fig. 9. The external voltage V_{ext} is the sum of the voltages in Fig. 9

$$V_{ext} = R_s I_{tot} + L\frac{dI_{tot}}{dt} + V \tag{18}$$

in combination with the total current I_{tot}

$$I_{tot} = I(V) + C\frac{dV}{dt} \tag{19}$$

leads to the differential equation for the equivalent circuit of the RTD:

$$\frac{d^2V}{dt^2} + \left(\frac{\frac{dI}{dV}(V)}{C} + \frac{R_s}{L}\right)\frac{dV}{dt} + \frac{R_s}{LC}I(V) + \frac{V}{LC} = \frac{V_{ext}}{LC}. \tag{20}$$

The differential equation is reduced to

$$\frac{d^2v}{d\tau^2} + (i'(v)r)\frac{dv}{d\tau} + ri(v) + v = 1 \tag{21}$$

by scaling the time as $t = \sqrt{LC}\ \tau$, the voltage at the diode as $V = vV_{ext}$, the series resistance as $R_s = \sqrt{L/C}\ r$, and the current through the diode as $I(V) = \sqrt{C/L}\ V_{ext}i(v)$. With $\omega = \frac{dv}{d\tau}$ it follows

$$\frac{d\omega}{d\tau} = -(i'(v) + r)\,\omega - r\,i(v) - v + 1. \tag{22}$$

Fixed points for the differential equation are given by:

$$i(v) = \frac{1-v}{r} \tag{23}$$

as is shown in Fig. 10. In this diagram the intrinsic I-V curve is plotted in combination with the I-V curve obtained from a series resistance at a certain supply voltage V_{ext}.

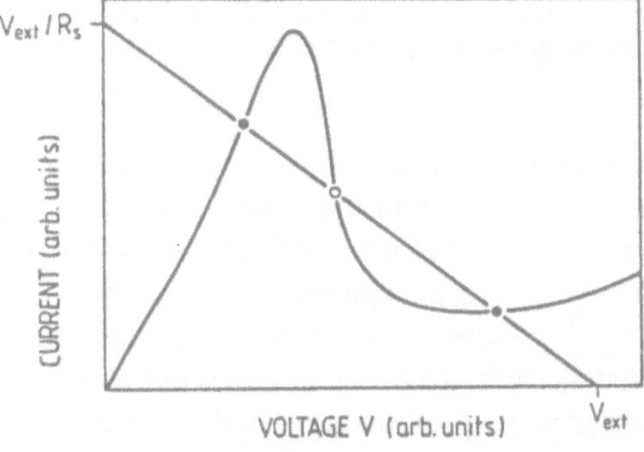

Figure 10
Loadline model for a resonant tunneling diode with an external series resistance. Crossing points of the I-V curve of the series resistance and the intrinsic current voltage characteristics of the resonant tunneling diode are the fixed points of the differential equation 22. Closed cycles are stable and open cycles are unstable fixed points.

The crossing points (cycles in the diagram) correspond to the fixed points of the differential equation (22). With this geometric solution of the transcendent Eq. (22) the allowed DC-points for the RTD at a certain supply voltage V_{ext} are found. But it is not directly apparent whether the DC-points found in this way are stable or not. To get information about the stability of the DC-points a linearization of the differential Eq. (22) around the fixed points (at v=v*) is necessary. This leads to the following differential equation system:

$$\begin{pmatrix} \frac{d\Delta v}{d\tau} \\ \frac{d\Delta w}{d\tau} \end{pmatrix} = \begin{pmatrix} 0 & 1 \\ -(1 + ri'(v^*)) & -(i'(v^*) + r) \end{pmatrix} \begin{pmatrix} \Delta v \\ \Delta w \end{pmatrix} \tag{24}$$

The solution follows as:

$$\begin{pmatrix} \Delta v \\ \Delta w \end{pmatrix} = \exp\left(\underline{\underline{A}}\tau\right) \begin{pmatrix} \Delta v_0 \\ \Delta w_0 \end{pmatrix} \tag{25}$$

The eigenvalues $\lambda_{1,2}$ of the matrix $\underline{\underline{A}}$ determine whether solutions at the fixed points are stable or not. Because of the exponential solution of the differential equation (24) stable fixed points can only be found in the regime where the real part of the eigenvalues are negative. The corresponding stability criteria are shown in Fig. 11. For a more detailed analysis of power and stability limitations of resonant tunneling diodes see also Kidner et al. [32].

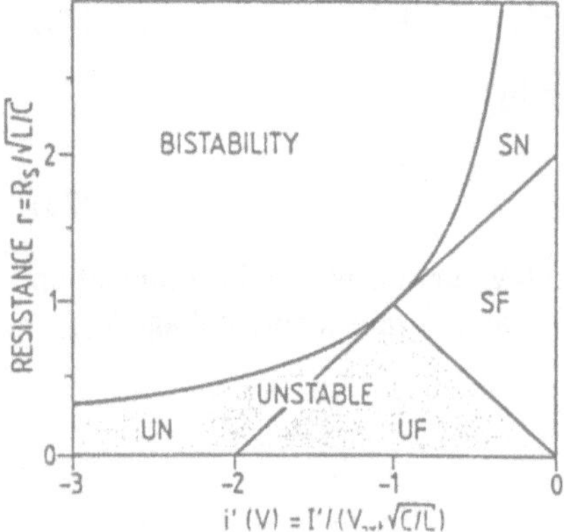

Figure 11
Stability analysis of fixed points of Fig. 10. The series resistance and the derivation of the current, $i'(V)$ is shown in units of $(\sqrt{L/C})$. UN, UF, SN and SF denote unstable node, unstable focus, stable node and stable focus, respectively. In the shaded area the RTD is unstable and oscillations are the result of the series resistance, the capacity and the series inductance. In the BISTABILITY (saddle point) area the I-V curve shows a hysteresis effect and the diode is not able to oscillate. (After Lange et al. [54])

According to the analysis there are two areas of instability in the r(i') diagram (shaded area in Fig. 11) where oscillations are generated in the circuit. It is directly apparent from Fig. 11 that not every fixed point in the negative differential resistance region ($i' < 0$) gives rise to oscillations. There is a critical dependence on the absolute value of the resistance with respect to $\sqrt{L/C}$. In some cases it should therefore be possible to stabilize the diode by a series resistance whose absolute value is in the stable node or stable focus regime. The most common way of stabilizing a diode is by means of a parallel capacitor. Because of the resistance scaling in units of $\sqrt{L/C}$ a capacity variation follows a hyperbolic curve in the diagram. Whether one can switch from a stable to an unstable area depends critically on the starting point in the diagram. If the fixed point is located in the bistable area (saddle point) one cannot leave this region by just changing the inductance or capacity of the measuring circuit. The bistability in this region is caused by a large series resistance.

Above a certain voltage drop across the resistance the I-V curve becomes bistable. Since the voltage drop across the resistance R_s is proportional to the absolute current, bistability in the case of large mesas (high total currents) appears at low series resistance. The "critical" resistance R_c at which bistabilities occur in the I-V curve can be estimated from the intrinsic curve as:

$$R_c = \frac{1}{A}\frac{U_v - U_p}{j_p - j_v} \quad (26)$$

(U_p, U_v, j_p and j_v denote peak and valley voltage and peak and valley current density, respectively). A reduction in the total current by a factor of 10 (using smaller mesas) means that the critical resistance can be enhanced by the same factor, which makes extrinsic bistabilities much more unlikely.

Besides extrinsic bistabilities, intrinsic bistabilities are also reported in the literature [33, 34]. Bistabilities in the I-V curve due to charge accumulation effects in the well are observed i. e. by Baba et al. [35].

4 Experimental results

4.1 The effect of the growth temperature on the performance of AlAs double barrier diodes embedded in GaAs

The first negative differential resistance (NDR) was observed at RTDs based on heterostructures by Chang et al. in 1974 [1]. Nevertheless it took nearly ten more years until the first reports about resonant tunneling diodes operating at room temperature appeared (see ref.[2]). This result coincides with the improvement in MBE techniques. The improvement in material and interface quality therefore seems to be the most likely reason for the improvement in RTD quality for that long time. Most of the resonant tunneling diodes reported in the literature are based on AlAs or AlGaAs barriers embedded in GaAs or on pseudomorphic InGaAs in the GaAs and InP system where they were grown by molecular beam epitaxy. While for AlAs or AlGaAs barriers embedded in GaAs the PVR does not exceed values much larger than 7 at room temperature. Broekaert et al. [36] and Smet et al. [37] reported a PVR of 30 and 50, respectively, at room temperature in $In_{0.53}Ga_{0.47}As$/AlAs/InAs RTDs. A small InAs layer in the well was used in these RTDs to reduce the resonance energy. Therefore the authors were able to reduce the well width in the structure leading to less scattering in the sample. Although the reason for the large difference in PVR compared to diodes in the GaAs system is not completely clear one can also think of less interface scattering in this kind of diodes.

Because of high current densities needed in applications of RTDs, the barriers are very small (several monolayers MLs) and the interface has to be very sharp. It was previously shown that AlAs/GaAs interfaces do not exhibit chemically abrupt, completely flat interfaces - even under optimum growth conditions [38]. The transition between GaAs and AlAs occurs within 2 and 4 monolayers. In addition, a lateral interface roughness is superimposed on the diffuse transition which can be characterized by step distances and step heights. Therefore the overall width of the interface region is determined by the width of the chemical transition in the growth direction and the superimposed lateral roughness. Both depend sensitively on growth temperature. As a consequence the interface quality is strongly dependent on the growth temperature and hence a significant effect of the substrate temperature on the performance on RTDs is expected.

A systematic study of the performance of resonant tunneling diodes as a function of substrate temperature has shown [39, 40, 16] that an optimum growth temperature for AlAs/GaAs diodes can be found. The peak-to-valley ratio (PVR), serving as a figure of merit for the quality of the diode, increases with substrate temperature up to the temperature range between 580°C or 600°C (Fig. 12).

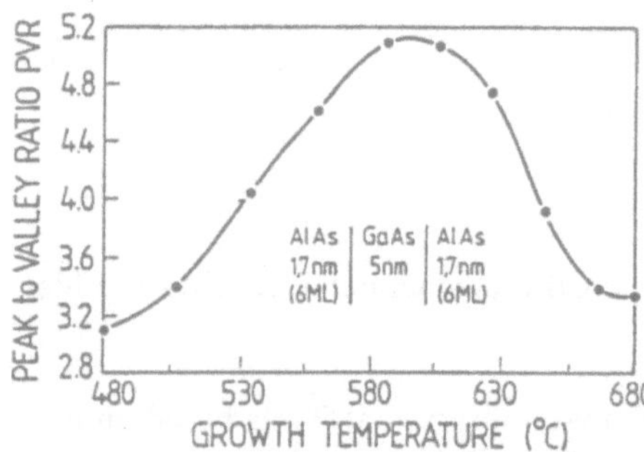

Figure 12
Room temperature current peak-to-valley ratio (PVR) of 6 monolayer (ML) AlAs double barrier diodes embedded in GaAs as a function of growth temperature. The well width is 5 nm. (After Förster et al. Ref. [39])

Above that temperature the PVR decreases rapidly. Reasons for this behavior can be found in the microstructure of the diodes. In Fig. 13 high resolution transmission electron microscopy (HRTEM) pictures of AlAs resonant tunneling diodes grown at different growth temperatures are depicted with atomic resolution.

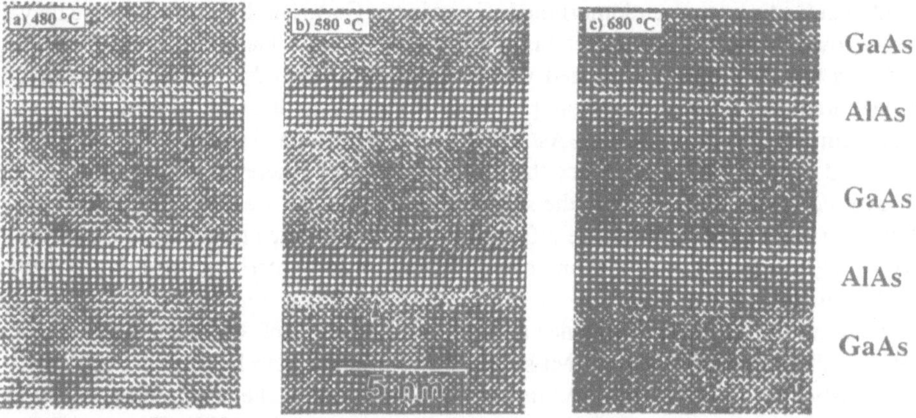

Figure 13 (100) HRTEM cross section images of 6 monolayer (ML) AlAs double barrier diodes embedded in GaAs grown at different growth temperatures: a) 480°C, b) 580°C and d) 680°C. (After Förster et al. Ref. [39])

At low substrate temperatures, the normal (AlAs on GaAs) and the inverted (GaAs on AlAs) interface exhibit similarly rough morphologies in terms of step distances. The interdiffusion of the group III element is small leading to narrow transition widths between GaAs and AlAs. Therefore, the RTD structure is symmetric because both interfaces have nearly the same microstructure. From the gradual change in the image patterns between GaAs and AlAs, a chemical transition width of 2 to 3 ML is determined. The most likely explanation for the diffuse appearance of the interface is that the scale of lateral roughness is smaller than the specimen thickness of 8 to 18 nm. This high density of lateral steps found in HRTEM probably causes scattering and leads to a lower PVR at low growth temperatures. The fact that interface roughness yields a relevant scattering mechanism has already been discussed in sec. 2.3. Simulations by Chevoir and Vinter [19] have shown that interface scattering decreases the PVR. Leo and McDonald calculated the effect of interface roughness as a function of the location of the interface and concluded that scattering effects at the interfaces adjacent to the well are especially important.

An abrupt change of the image patterns within 1 - 2 MLs between GaAs and AlAs is found at samples grown at the optimum growth temperature in the temperature range between 580°C and 600°C. Monolayer steps can be observed at the normal interface (indicated by an arrow in Fig. 13b) showing that the lateral step distance is significantly enhanced compared to samples grown at low growth temperatures. Interfaces are optimized and scattering effects are reduced. This high interface quality could be the reason for a result of Reddy et al. [41], who did not find any difference in PVR with and without growth interruption at RTDs grown at 600°C. They concluded from this finding that interface roughness at inverted and normal interfaces does not play a significant role in determining the perpendicular transport characteristics of DBRTDs.

The HRTEM picture of diodes grown at 600°C furthermore shows that the top barrier is slightly (one monolayer) enhanced in thickness with respect to the bottom barrier. For this structure the highest PVR was 5.35 at room temperature. Similar asymmetries are also found by Tsao et al. [42]. They have shown that a slightly higher PVR can be observed at intentionally asymmetric diodes. The highest PVR (PVR = 5.6) was found at AlAs double barriers with a nominally 6 ML bottom and a 7 ML thick top barrier.

For very high substrate temperatures like 680°C the barrier thicknesses are increased (AlGaAs rather then pure AlAs) and the interface quality is reduced resulting in a lower PVR.

A strong temperature dependence is not only found in the PVR but also in the symmetry of the I-V characteristics of the diode. The ratios of the supply voltages at peak maximum and the corresponding peak-to-valley ratios for positive and negative supply voltages are chosen as parameters to demonstrate the degree of asymmetry in Fig. 14.

RTDs grown at substrate temperature up to 530°C are quite symmetric both in the peak voltages and in the PVRs. At substrate temperatures between 530°C and 630°C a slight asymmetry is observed whereas above 630°C a dramatic increase in asymmetry

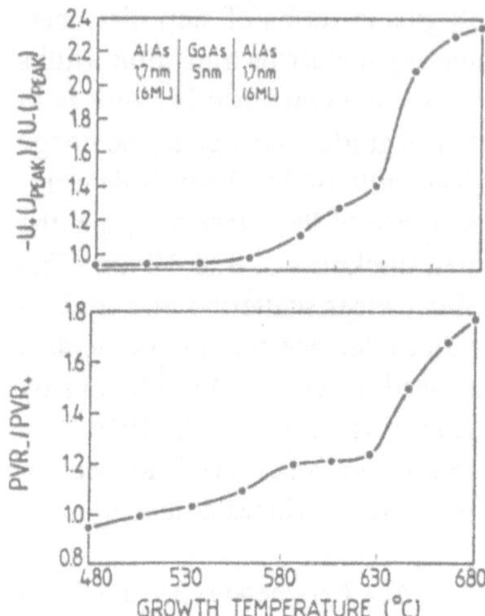

Figure 14
Ratio of the peak-voltages (U_{p-}/U_{p+}), a) and the peak-to valley ratios (PVR_-/PVR_+), b) in the positive and negative bias voltage direction, supplied to the top layer of 6 monolayer double barrier resonant tunneling diodes with a 5 nm GaAs well. (After Förster et al. Ref. [39])

as a result of silicon doping segregation towards or into the double barrier region is found. This effect is shown in the electron concentration profile of samples grown at 480°C, 580°C and 680°C (Fig. 15). The corresponding layer sequence is as follows: 1μm n_+ - GaAs buffer ($n_+ = 5 \times 10^{18}cm^{-3}$), 10 nm n GaAs ($n = 10^{17}cm^{-3}$), 10 nm n_- GaAs ($n_- = 10^{16}cm^{-3}$), 5 nm GaAs, 1.7 nm AlAs, 5 nm GaAs, 1.7 nm AlAs, 5 nm GaAs, 10 nm n_- GaAs, 10 nm n GaAs, 0.5 μm n_+ GaAs.

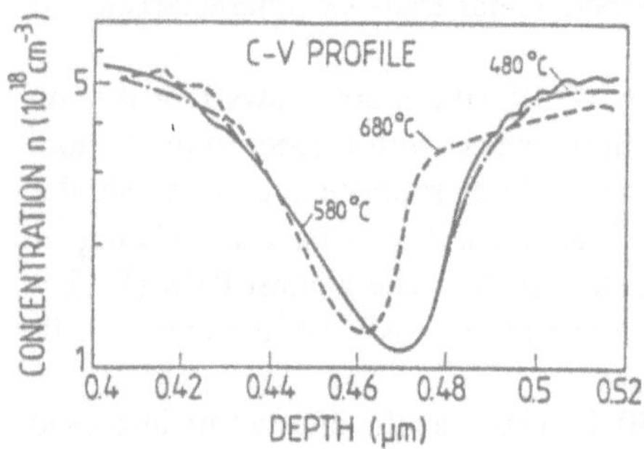

Figure 15
Electron concentrations as a function of depth within the sample obtained from capacitance voltage measurements for AlAs double barrier diodes embedded in GaAs grown at different growth temperatures. (After Förster Ref. [39])

While the profiles for samples grown at 480°C and 580°C show the same spatial dependence, the profile for the sample grown at 680°C exhibits a clearly decreased nominally undoped region shifted towards the growth direction. This effect tilts the potential diagram of the double barrier region and leads to the observed asymmetry.

4.2 The effect of scattering on the room temperature PVR of AlAs double barrier diodes

As discussed above scattering-related reduction of the PVR can be minimized for 6 ML AlAs double barrier diodes in GaAs by growing at the optimum growth temperature between 580°C and 600°C. Nevertheless the maximum PVR one can achieve in AlAs double barrier diodes based on GaAs is only slightly above 5 at room temperature. To overcome these limitations the incorporation of a pseudomorphic emitter-side InGaAs quantum well improve both the PVR and the current density j_p. Brugger et al. [27] have shown that a 4 nm $In_{0.15}Ga_{0.85}As$ in front of a $Al_{0.7}Ga_{0.3}As/GaAs/Al_{0.7}Ga_{0.3}As$ double barrier structure yields PVR = 7.2 (28) at 300 K (77 K), which are the highest reported values in the AlGaAs/GaAs system.

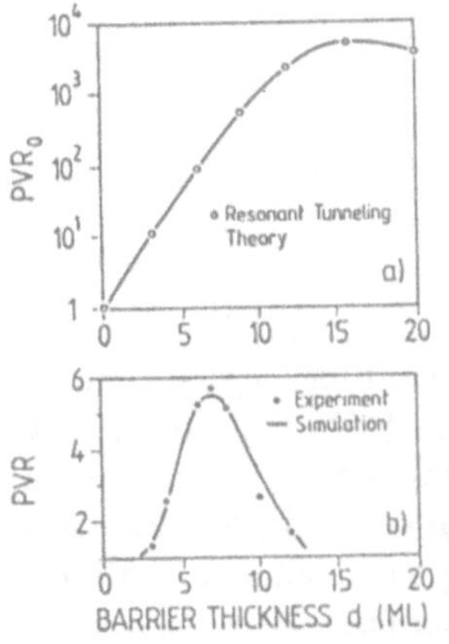

Figure 16
a) Logarithmic plot of the peak-to-valley ratio (PVR) of theoretically simulated characteristics of an RTD consisting of AlAs barriers with varying thickness d (in monolayers, ML), embedded in $In_{0.17}Ga_{0.83}As$ with well width of 5 nm. Only resonant tunneling is taken into account. (b) PVR measured on RTDs (dark circles) consisting of two AlAs barriers with varying thickness d, embedded within $In_{0.17}Ga_{0.83}As$. The area of the diodes is $7 \times 7\mu m^2$. The simulation (curve in solid line) includes resonant tunneling as well as sequential tunneling.

A systematic study of AlAs barrier thickness variation is shown in Fig. 16b for the pseudomorphic $In_{0.17}Ga_{0.83}As$ system after Lange et al. [44]. It was found that for barrier thicknesses larger than 7 ML the PVR is strongly reduced. One can clearly observe a maximum in PVR for diodes with barrier thicknesses of about 7 MLs at room temperature. The same tendency was also found for AlAs barriers embedded in GaAs at room temperature and even at 77 K with maximum PVR-values of 12 [43].

In order to understand the overall dependence of the PVR on barrier thickness, I-V curve simulations were performed using the transfer matrix formalism for the coherent tunneling process through both barriers (Fig. 16a) and taking scattering into account by means of Eq. (17).

As is shown in Fig. 16a the PVR for coherent tunneling without scattering increases very strongly up to barrier thicknesses of about 15 ML. Above that value one can see the influence of thermal emission through the second quasi-bound state in the well. The peak current density affected by the half width of the first resonance in the transmission probability decreases very rapidly with barrier thickness (see Eq. (11)) while the current passing the higher quasi bound state decreases more slowly with barrier thickness. This effect causes the decay in PVR above 15 ML.

In simulations, where only coherent tunneling is taken into account, PVRs are much higher than in the experimental results. As is shown in sec. 2.3 inelastic scattering can be included by folding the transmission probability with a normalized Lorentzian of half width Γ_i. By taking into account only a constant scattering contribution, simulations roughly exhibit the tendency with a maximum PVR at about 6 ML. The decrease in PVR for higher barriers can be understood in terms of resonance broadening according to the discussion in sec. 2.3. The higher the barrier thickness the smaller is the resonance width leading to a high PVR suppression in the diode (see Eq. (17)).This effect therefore leads especially for large barrier thicknesses to a very high suppression in the PVR at room temperature.

Furthermore the strong decrease in PVR for small barriers (smaller than 6 ML) can be explained by interface roughness effects. It is shown by Lange et al. [44] that even at room temeprature interface roughness can be a dominant mechanism for the PVR reduction in RTDs with very thin barriers (smaller than 6 ML).

5 Applications

The practical importance of resonant tunneling diodes has meanwhile been proven by a variety of applications. In 1984 oscillations were observed for the first time on double barrier resonant tunneling structures by Sollner et al. [45]. A coaxial cavity was used as a resonant circuit for oscillating quantum well resonators. The maximum oscillation frequency was 18 GHz and the maximum output power 5 μW. Electrical connection was made to the ohmic contact in such devices by a pointed wire (whisker) direct on top of the mesa whose diameter was only a few microns.

Meanwhile frequencies of several hundreds of GHz are obtained in rectangular waveguide resonators using whisker contacts . Brown et al. [46] compare AlAs/GaAs and InAs/AlSb RTDs in a waveguide resonator and find oscillation frequencies as high as 712 GHz with a power of 0.3 μW with InAs/AlSb diodes (Fig. 17).

The maximum power density of $90W/\mathrm{cm}^{-2}$ achieved with a 1.8 μm diameter InAs/AlSb RTD at 350 GHz was 50 times higher than for the AlAs/GaAs system. This finding is probably related to advantages of InAs/AlSb diodes over GaAs/AlAs diodes for high frequency applications. The main reasons might be the higher band discontinuity (higher current density at the same PVR) leading to a reduced RC time delay in the RTDs, the higher drift velocity in InAs than in GaAs and the lower series resistance R_s caused by the higher electron mobility in InAs. The maximum achievable frequency in a RTD resonator can be estimated as

$$f_{max} = \frac{1}{2\pi C}\sqrt{\frac{-G_{max}}{R_s} - G_{max}^2} \tag{27}$$

(see Sollner et al. [47]) where C is the capacity of the diode and G_{max} the maximum negative value of dynamic conductance in the NDR region of the I-V curve. The corresponding maximum power is estimated as: $P_{max} = \frac{3}{16}\Delta I \Delta V$ (see ref. in Sollner et al. [45]), where ΔI and ΔV is the difference in peak and valley current and voltage, respectively. Therefore the power increases with increasing mesa size. On the other hand the mesa size is limited due to the occurrence of bistabilities at a certain series resistance as discussed in Sec. 3. Nevertheless small mesas are needed to achieve high frequencies (low diode capacity) and therefore the power is usually limited within the μW region.

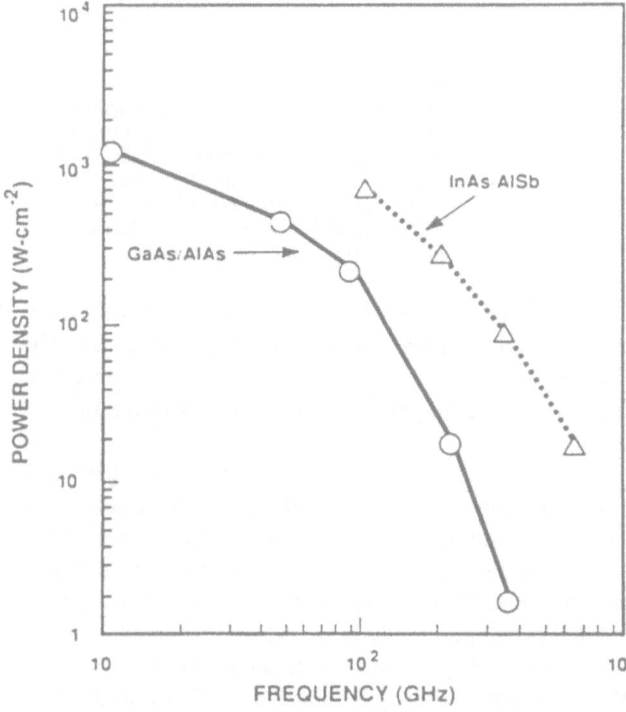

Figure 17
Experimental oscillator results for the GaAs/AlAs RTD (solid line) and the best InAs/AlSb diodes (dashed line). All oscillations were observed at room temperature. (After Brown et al. Ref. [46])

The high nonlinearity of RTD I-V curves and the usefulness of RTDs as oscillators can be combined in a self-oscillating mixer. This is shown e. g. by Millington et al. [48], who built a 11 GHz self-oscillating microwave mixer using a AlGaAs/GaAs double barrier RTD. The authors demonstrated that such RTD mixers yield the same performance as conventional mixers but with considerably less complexity.

Furthermore, high frequencies can be generated with RTDs by frequency multiplying. Supply of a sinusoidal voltage to an unbiased RTD with a voltage amplitude higher than the peak voltage in the diode I-V characteristics causes at least three maxima over one cycle. This is shown by Sollner et al. [47]. Hence a RTD can be used as a very efficient odd-harmonic generator.

An odd-harmonic multiplier for the W-band based on an AlAs double barrier diode with AlAs barriers and an AlGaAs prebarrier is presented by Brugger et al. [49] (see Fig. 18).

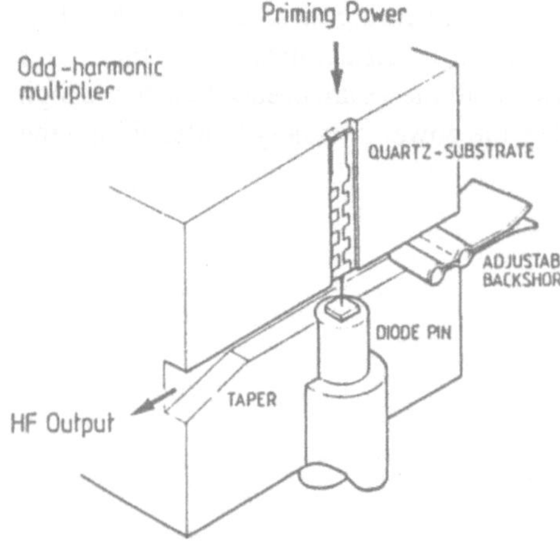

Figure 18
Scheme of an odd-harmonic multiplier based on AlAs double barrier resonant tunneling diodes. Input power 0 dbm at 23.8 - 26.4 GHz injected in an 8 μm diameter mesa. Output power -20 dbm (10 μW) at 71.4 - 79.2 GHz. (Courtesy of H. Brugger and J. Schroth, Daimler Benz, see also Ref. [49])

A current density of 2×10^4 A/cm^2 and a PVR of 5 are reached. An input power of 0 dBm at 23.8 to 26.4 GHz is injected via a bias line to a 8μm diameter mesa. The sample was contacted by a whisker directly on top of the mesa. The output power of -20 dBm (10 μW) was measured at room temperature in the frequency range of 71.4 to 79.2 GHz.

A number of further applications are also discussed by Capasso et al. [50] and Luryi [51]. The authors have demonstrated the functionality of resonant tunneling bipolar transistors. It was shown that those devices can be used as oscillators with a tuneable frequency range. Furthermore the bistable character of RTDs with multiple peak characteristics were used for logic devices. As an example a four-bit parity generator was demonstrated. All these experiments have shown that there is great potential for resonant tunneling diodes in high frequency applications. The fact that the intrinsic response time in RTDs, first measured by Sollner et al. [52], lies in the THz region promises further very interesting applications for resonant tunneling diodes.

6 Concluding Remarks

The present paper demonstrates that resonant tunneling diodes are quantum mechanical devices which conceal a lot of interesting basic physical phenomena. Especially the improvement of MBE in the last ten years has opened a large potential for future applications. In particular the field of high frequency applications might benefit from these developments. But these might also be of cosiderable interest in basic research.

One can think of RTDs as detectors for scattering mechanisms or interface roughness in different material systems.

Besides the classical double barrier diodes, resonant interband tunneling diodes on the basis of InAlAs/InGaAs also seem to be very promising devices for application. Results from Day et al. [53], who reported a room temperature PVR of over 100 on these diodes, demonstrate very clearly the great improvement in tunneling diodes over the last 10 years. Further research in this field is therefore highly desirable.

Acknowledgements

I would like to thank Jens Lange for his helpful co-operation in this field and H. Lüth for a critical reading of the manuscript.

Bibliography

[1] L. L. Chang, L. Esaki and R. Tsu, *"Resonant tunneling in semiconductor double barriers"*, Appl. Phys. Lett. *24*, 593 (1974)

[2] Mashahiro Tsuchiya, Hiroyuki Sakaki and Junji Yoshino, *"Room Temperature Observation of Differential Negative Resistance in an AlAs/GaAs/AlAs Resonant Tunneling Diode"*, Jpn. J. Appl. Phys.*24*, L466 (1985)

[3] E. R. Brown, *"Resonant Tunneling in High-Speed Double Barrier Diodes"* in: *"Hot Carriers in Semiconductor Nanostructures, Physics and Applications"*, Jagdeep Shah, ed., p. 469, Academic Press, 1992, ISBN 0-12-6381

[4] B. Ricco and M. Ya. Azbel, *"Physics of resonant tunneling. The one-dimensional double-barrier case"* Phys Rev. *B 29*, 1970 (1984)

[5] M. Ya Azbel, *"Eigenstates and properties of random systems in one dimension at zero temperature"*, Phys Rev. *B 28*, 4106 (1983)

[6] Yuming Hu and Shawn P. Stapleton, *"Double-Barrier Resonant Tunneling Transport Model"*, IEEE Journal of Quantum Electronics. *29* (2), 327 (1993)

[7] G. A. Toombs and F. W. Sheard, *"The Background to Resonant Tunneling Theory"* in *"Electronic Properties of Multilayers and Low-Dimensional Semiconductor Structures"*, J. M. Chamberlain, L. Eaves and J. C. Portal, ed., Plenum Press New York and London, NATO ASI Series Volume *B 231*, p. 257 (1990) ISBN 0-306-43662-0

[8] E. E. Mendez, *"Physics of Resonant Tunneling in Semiconductors"* in *"Physics and Applications of Quantum Wells and Superlattices"*, E. E. Mendez and K. von Klitzing, ed., Plenum Press New York and London, NATO ASI Series Volume *B 170*, p. 159 (1987) ISBN 0-306-42823-7

[9] R. Tsu and L. Esaki, *"Tunneling in a finite superlattice"*, Appl. Phys. Lett. *22* (11), 568 (1973)

[10] T. Weil and B. Vinter, *"The equivalence between resonant and sequential tunneling on double-barrier diodes"*, Appl. Phys. Lett. *50* (18), 1281 (1987)

[11] Peter J. Price, *"Theory of resonant tunneling in heterostructures"*, Phys. Rev. *B 38*, 1994 (1988)

[12] Serge Luryi, *"Frequency limit of double-barrier resonant tunneling oscillators"*, Appl. Phys. Lett. *47* (5), 490 (1985)

[13] H. Morcoc, J. Chen, U. K. Reddy, and T. Henderson, *"Observation of negative differential resistance due to tunneling through a single barrier into a quantum well"*, Appl. Phys. Lett. *49* (2), 70 (1986)

[14] M. Büttiker, *"Coherent and sequential tunneling in series barriers"*, IBM J. Res. Develop. *32* (1), 63 (1988)

[15] A. Douglas Stone and P. A. Lee, *"Effect of Inelastic Processes on Resonant Tunneling in One Dimension"*, Phys Rev. Letters *54* (11), 1196 (1985)

[16] A. Förster, J. Lange, D. Gerthsen, Ch. Dieker and H. Lüth, *"The Effect of Interface Roughness and Scattering on the Performance of AlAs/InGaAs Resonant Tunneling Structures"*, J. Vac. Sci. Technol. *B 11* (4), 1743 (1993)

[17] Jyh-Haur Tyan, Juh-Tzeng Lue, and Jian-Shang Sheng, J., *"Effects of inelastic electron-phonon scattering on the resonant tunneling in double-barrier structure"*, Appl. Phys. Lett. *68* (6), 2829 (1990)

[18] V. J. Goldman, D. C. Tsui, and J. E. Cunningham, *"Evidence for LO-phonon-emission-assisted tunneling in double-barrier heterostructures"*, Phys. Rev. *B 36*, 7635 (1987)

[19] Francois Chevoir and Borge Vinter, *"Scattering-assisted tunneling in double-barrier diodes: Scattering rates and valley current"*, Phys. Rev. *B 47* (12), 7260 (1993)

[20] F. Chevoir and B. Vinter, *"Calculation of phonon-assisted tunneling and valley current in a double barrier diode"*, Appl. Phys. Lett. *55* (18), 1859 (1989)

[21] P. Guéret, C. Rossel, W. Schlup, and H. P. Meier, *"Investigations on resonant tunneling in III-V heterostructures: Comparison between data and model calculations"*, J. Appl. Phys. *66* (9), 4312 (1989)

[22] James Leo and A. H. McDonald, *"Disorder-Assisted Tunneling through a Double-Barrier Structure."*, Phys. Rev. Lett. *64* (8), 817 (1990)

[23] E. E. Mendez, E. Calleja, C. E. T. Goncalves da Silva, L. L. Chang *"Observation by resonant tunneling of high-energy states in GaAs-$Ga_{1-x}Al_xAs$ quantum wells"*, Phys. Rev. *B 33* (10), 7368 (1986)

[24] E. E. Mendez, W. I. Wang, E. Calleja, and C. E. T. Goncalves da Sil *"Resonant tunneling via X-point states in AlAs-GaAs-AlAs heterostructures"*, Appl. Phys. Lett. *50* (18), 1263 (1987)

[25] E. E. Mendez, E. Calleja and W. I. Wang, *"Tunneling through indirect-gap semiconductor barriers"*, Phys. Rev. *B 34* (8), 6026 (1986)

[26] Sadao Adachi, *"GaAs, AlAs, and $Al_xGa_{1-x}As$: Material parameters for use in research and device applications"*, J. Appl. Phys. *58* (3), R1 (1985)

[27] H. Brugger, U. Meiners, C. Wölk, R. Deufel, A. Marten, M. Rossmanith K. v. Klitzing, and R. Sauer, *"Pseudomorphic Two-Dimensional Electron-Gas-Emitter Resonant Tunneling Devices"*, Proceedings of the 21st European Solid State Device Research Conference, ESSDERC '91, Lausanne, Microelectronics Engineering *15*, 663 (1991)

[28] R. Diniz, J. Smoliner and E. Gornik, U. Meiners, H. Brugger, P. Wisniewski and T. Suski *"Hydrostatic pressure studies of thin barrier resonant tunneling diodes"* Accepted for publication in Semiconductor Science and Technology, to appear in 1993

[29] H. Brugger, U. Meiners, R. Diniz, T. Suski, E. Gornik, A. Förster, *"Hydrostatic Pressure Sensor based on Solid State Tunneling Devices"*, Proceedings of the 6th International Conference on Modulated Semiconductor Structures MSS-6, Garmisch-Partenkirchen (Germany), Aug. 23-27, 1993, appears in Solid State Electronics

[30] Jeff F. Young, B. M. Wood, H. C. Liu, M. Buchanan, D. Landheer, A. J. Spring Thorpe and P. Mandeville, *"Effect of circuit oscillations on the dc current-voltage characteristics double barrier resonant tunneling structures"*, Appl. Phys. Lett. *52* (17), 1398 (1988)

[31] O. H. Huges, M. Henini, E. S. Alves, L. Eaves, M. L. Leadbeater, T. J. Foster, F. W. Sheard, G. A. Toombs, A. Celeste and J. C. Portal, *"Investigation of double barrier resonant tunneling devices based on (AlGa)As/GaAs"*, J. Vac. Sci. Technol. *B 6* (4), 1161 (1988)

[32] C. Kidner, I. Mehdi, J. R. East and G. I. Haddad, *"Power and Stability Limitations of Resonant Tunneling Diodes"*, IEEE Transactions on Microwave Theory and Techniques, *38* (7) 864 (1990)

[33] A. Zaslavsky, V. J. Goldman, D. C. Tsui, J. E. Cunningham, *"Resonant tunneling and intrinsic bistability in asymmetric double barrier heterostructures"*, Appl. Phys. Lett. *53* (15), 1408 (1988)

[34] V. J. Goldman, D. C. Tsui and J. E. Cunningham, *"Observation of Intrinsic Bistability in Resonant-Tunneling Structures"*, Phys. Rev. Letters *58* (12), 1256 (1987)

[35] T. Baba and M. Mizuta, *"Existence of intrinsic bistability in resonant tunneling diode studied by novel Monte Carlo technique"* in *"Gallium arsenide and related compounds 1989"*, International Symposium on Gallium Arsenide and Related Compounds 16: Proceedings, Karuizawa, Japan , IOP Publ. Bristol, UK (1990), p. 807

[36] Tom P. E. Broekaert, Wai Lee, and Clifton G. Fonstad, *"Pseudomorphic $In_{0.53}Ga_{0.47}As/AlAs/InAs$ resonant tunneling diodes with peak-to-valley current ratios of 30 at room temperature"*, Appl. Phys. Lett. *53* (16) 1545 (1988)

[37] Jurgen H. Smet, Tom P. E. Broekaert, and Clifton G. Fonstad, *"Peak-to-valley ratios as high as 50:1 at room temperature in pseudomorphic $In_{0.53}Ga_{0.47}As/AlAs/InAs$ resonant tunneling diodes"*, J. Appl. Phys. *71* (5) 2475 (1992)

[38] D. Bimberg, F. Heinrichsdorff, R. K. Bauer, D. Gerthsen, D. Stenkam, D. E. Mars and J. N. Miller, *"Binary AlAs/GaAs versus ternary GaAlAs/GaAs interface: A dramatic difference of perfection"*, J. Vac Sci. Technol. *B 10* (4), 1793 (1992)

[39] A. Förster, J. Lange, D. Gerthsen, Ch. Dieker and H. Lüth, *"The Effect of Growth Temperature on AlAs/GaAs Resonant Tunneling Diodes"*, to appear in J. Physics D: Applied Physics.

[40] Ch. Dieker, D. Gerthsen, A. Förster, J. Lange and H. Lüth, *"Microstructure of GaAs/AlAs and InGaAs/AlAs resonant tunneling diodes and its correlation with the electrical properties"*, to be published in Institute of Physics Conference Series 1993

[41] V. K. Reddy and D. P. Neikirk, *"Influence of growth interruption on I-V characteristics of AlAs/GaAs double barrier resonant tunneling diodes"*, J. Vac. Sci. Technol. *B 10* (2), 1045 (1992)

[42] A. J. Tsao, V. K. Reddy, D. R. Miller, K. K. Gullapalli, and D. P. Neikirk, *"Effect of barrier thickness asymmetries on the electrical characteristics of AlAs/GaAs double barrier resonant tunneling diodes"*, J. Vac Sci. Technol. *B 10* (2), 1042 (1992)

[43] H. Brugger, private communication

[44] J. Lange, A. Förster, H. Lüth, and H. Brugger, *"Resonant and sequential tunneling in AlAs double barrier resonant tunneling diodes"*, to be published.

[45] T. C. L. G. Sollner, P. E. Tannenwald, D. D. Peck, and W. D. Goodhue *"Quantum well oscillators"*, Appl. Phys. Lett. *45* (12), 1319 (1984)

[46] E. R. Brown, J. R. Söderström, C. D. Parker, L. J. Mahoney, K. M. Molvar and T. C. McGill, *"Oscillations up to 712 GHz in AlAs/AlSb resonant-tunneling diodes"*, Appl. Phys. Lett. *58* (20), 2291 (1991)

[47] T. C. Sollner, E. Brown, W. D. Goodhue, and H. Q. Le, *"Microwave and Millimeter-Wave Resonant Tunneling Devices"* in *"Physics of quantum electron devices"*, Federico Capasso (Ed.), Springer Series in Electronics and Photonics 28, 1990 ISBN 3-540-51128-8 Springer-Verlag Berlin Heidelberg New York

[48] Garry Millington, Robert E. Miles, Roger D. Pollard, D. Paul Steenson and J. Martyn Chamberlain, *"A Resonant Tunnelling Diode Self-Oscillating Mixer with Conversion Gain"*, IEEE Microwave and Guided Wave Letters, *1* (11), 3204 (1991)

[49] H. Brugger, U. Meiners, C. Wölk, R. Deufel, J. Schroth, A. Förster *"High Quality GaAs-Based Resonant Tunneling Diodes For High Frequency Device Applications"*, Proceedings of the 13th IEEE/Cornell Conference on Advanced Concepts in High Speed Semiconductor Devices and Circuits, Cornell University, Ithaca, N. Y. (USA), August 5-7, 39 (1991), IEEE, ISBN No. 0-7803-0491-8

[50] F. Capasso, S. Sen, F. Beltram, and A. Y. Cho, *"Resonant Tunnelling and Superlattice Devices: Physics and Circuits"* in *"Physics of quantum electron devices"*, Federico Capasso (Ed.), Springer Series in Electronics and Photonics 28, 1990 ISBN 3-540-51128-8 Springer-Verlag Berlin Heidelberg New York

[51] Serge Luryi, *"Electronic Devices Using Multilayer Structures"* in *"Physics, Fabrication, and Applications of Multilayered Structures"*, P. Dhez and C. Weisbuch, ed. , Plenum Press New York and London, NATO ASI Series, Volume B182 p. 241 (1988), ISBN 0-306-42995-0

[52] T. C. L. G. Sollner, W. D. Goodhue, P. E. Tannenwald, C. D. Parker, *"Resonant tunneling through quantum wells at frequencies up to 2.5 THz"*, Appl. Phys. Lett. *43* (6), 588 (1983)

[53] D. J. Day, Rui Q. Yang, Jian Lu, and J. M. Xu, *"Experimental demonstration of resonant interband tunnel diode with room temperature peak-to-valley current ratio over 100."*, J. Appl. Phys. *73* (3), 1542 (1993)

[54] J. Lange, Diploma Thesis 1993, RWTH Aachen, Germany

Reproducible Quantum Conductance Fluctuations in Disordered Systems

Bernhard Kramer

Physikalisch-Technische Bundesanstalt Braunschweig, Bundesallee 100
3300 Braunschweig, Federal Republic of Germany

Summary: Experimental and theoretical results concerning the statistical properties of the conductance of electron systems at very low temperatures are reported. In the metallic regime the conductance fluctuations are of the order of e^2/h, independent of the size, and other parameters that characterize the system. The magnitude of these *universal conductance fluctuations* are determined by symmetry properties. In the insulating regime the configurational average of the conductance decreases much faster with the size of the system than its fluctuations such that the latter dominate transport in the thermodynamic limit at sufficiently low temperatures. The relation between the most probable value of the conductance, its configurational average, and the experimentally measured value are discussed. Conditions for the experimental observability of the fluctuations are briefly reviewed.

1 Introduction

In conventional statistical physics macroscopic physical quantities are identified with averages over statistical ensembles. The latter consist of many 'macroscopically equivalent' systems that are characterized by a certain macroscopically large number of independent degrees of freedom. Due to the independence of the latter statistical fluctuations within an ensemble are usually very small, relative to the average value, when the number of the degrees of freedom tend to infinity. The statistical averages obey the 'law of large numbers'. This can be most easily seen by analogy with the standard random walk problem. Ensemble averages are therefore the quantities to be identified with experimental results.

In recent years, with the availability of millikelvin temperatures in the laboratories [1], and the advent of preparation techniques that made the fabrication of sub-micron systems possible [2], quantum phenomena in the electronic properties became experimentally accessible that do not necessarily fulfil the classical 'law of large numbers'. The most prominent examples—to be discussed below—were found in electronic transport. All of the phenomena are related to the presence of *coherence of the quantum states* within macroscopically large regions of space [3] caused by the absence of inelastic, phase randomizing scattering events, for instance due to electron-phonon, and/or electron-electron interaction, at very low temperatures. They provide therefore insight into the coherent motion of quantum mechanical particles in a disordered, geometrically confined sample over macroscopic distances.

New approaches for the calculation of the electronic properties, including diagrammatic perturbational approximations [4], random matrix theories [5], as well as recursive numerical techniques [6], for electronic transport had to be designed in order to treat the quantum effects adequately. The question of the identification between theoretical and experimental 'physical' quantities had to be answered. Though many experimental and theoretical investigations of reproducible conductance fluctuations were performed during the past decade, the theoretical understanding is still incomplete. This is especially true for the insulating regime, were also the experiments are forbiddingly difficult. Here, hopping is dominating transport, and in addition to the statistics induced by quantum coherence the statistics of the hopping processes have to be taken into account.

The following chapters review the present status of experiments and theory, and provide a qualitative understanding of the phenomena.

2 Overview of Reproducible Conductance Fluctuations

2.1 Discovery

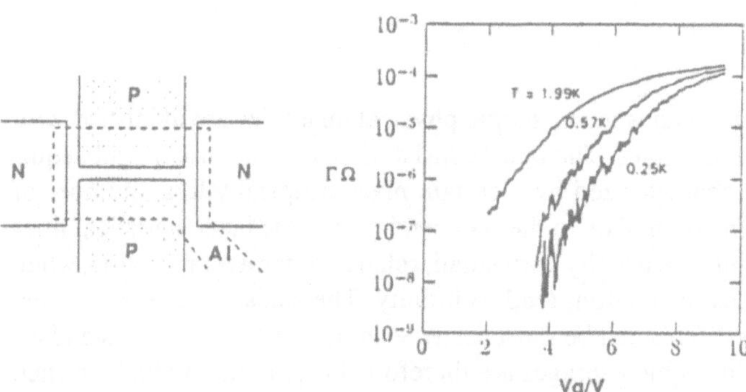

Figure 1
The reproducible fluctuations of the conductance Γ as observed at low temperatures in a quasi-one dimensional inversion layer channel when changing the voltage V_g applied to the Al gate on top of a Si MOSFET as shown schematically on the left (redrawn after [7]).

In the early eighties a group of researchers at IBM Yorktown Heights J. T. Watson Research Laboratory discovered a remarkable phenomenon when investigating the conductance of quasi-one dimensional inversion layers in Silicon MOSFETs at very low temperatures [7] (Fig. 1). When the temperature was sufficiently low, say below 100mK, the conductance exhibited stochastic fluctuations when the density of the charge carriers was changed by varying the gate voltage. They were *perfectly reproducible* for a given sample under given experimental conditions. The pattern of the fluctuations was, however, completely changed when the sample was heated up and cooled down again. Figure 2 shows an example that was obtained from a semiconductor-based sample, namely a SiGaAs wire [8].

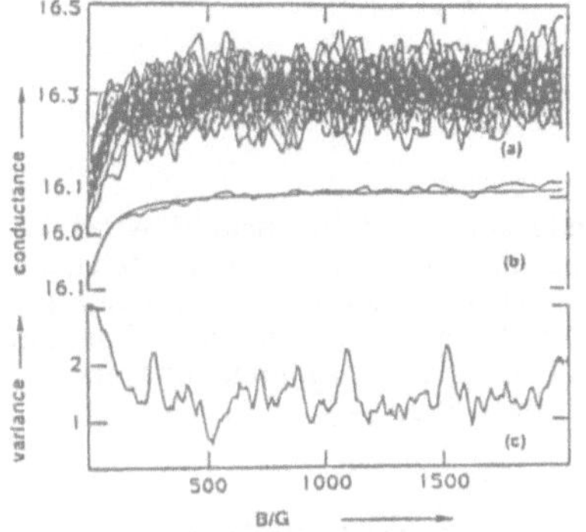

Figure 2
(a) Reproducible fluctuations of the magneto-conductance in units of e^2/h at T =45mK of a Si-GaAs wire after heating and cooling down again (46 cycles). (b) The mean and a one dimensional weak localization fit (phase coherence length L_Φ =3μm). (c) The variance of the 46 curves in (a) in units of $10^{-3}(e^2/h)^2$ (redrawn after [8]).

These, and similar fluctuations in thin metallic wires subject to a magnetic field, were thought to be related to the presence of randomness due to impurities and imperfections. Their dependence on the parameters that characterize the system provides not only insight into the nature of the quantum processes in quantum coherent systems with many degrees of freedom [9] but allow also for systematic quantitative studies of the statistical properties of non-self averaging quantities [10], and of the quantum properties of systems that exhibit chaotic behaviour in the classical limit [11, 12]. As a consequence, they were studied intensively during the past decade, experimentally [13] as well as theoretically [14]. A saturation of the activities is not yet in sight, especially since recent experiments seem also to indicate the importance of Coulomb effects [15].

2.2 Universality

In the metallic regime there is strong experimental evidence that the reproducible fluctuations are of the order e^2/h, and universal [8, 13, 16, 17, 18, 19, 20, 21]. They do not show a strong dependence on the degree of disorder, and the temperature, once it is below a certain critical value, and only a weak dependence on the shape of the sample (Fig. 3). However, there is experimental evidence that the magnitude of the universal conductance fluctuations depends in a characteristic way on fundamental symmetry properties [22]. If, for instance, a magnetic field applied to a system with no spin-orbit interaction is increased above certain critical values, say B_{c1} and B_{c2}, the fluctuations decrease by factors of two due to the cross-over from orthogonal (time reversal invariance) to unitary (no time reversal invariance) symmetry, and the Zeeman-splitting induced lifting of the spin degeneracy, respectively (Fig. 4). Similar characteristic behaviour is also expected for systems with spin-orbit coupling, i. e. symplectic symmetry [23, 24]. In addition, there is a theoretical prediction that the

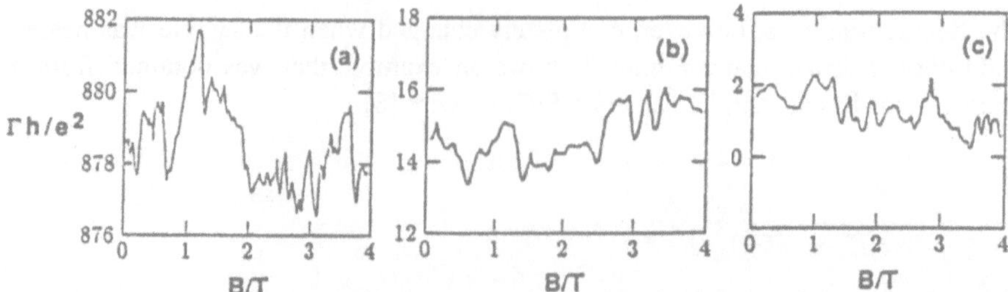

Figure 3 The universal fluctuations of the magneto-conductance Γ of thin metallic wires. (a) Mesoscopic Au ring of about 0.8μm diameter, (b) quasi-one dimensional inversion layer channel in a Si MOSFET and (c) numerical results for the disordered Anderson model. Although the three *samples* are microscopically (and macroscopically) different, as one can judge from the fact that the zero-magnetic field conductances are by about an order of magnitude different, the fluctuations are about of the same order of magnitude (redrawn after [14]).

universal conductance fluctuations depend on the coupling to the leads for sufficiently short system length [25]. Thus, the fluctuations in the metallic limit, and their characteristic properties, seem to be very well theoretically understood. Theoretical and experimental results do compare even on a quantitative level.

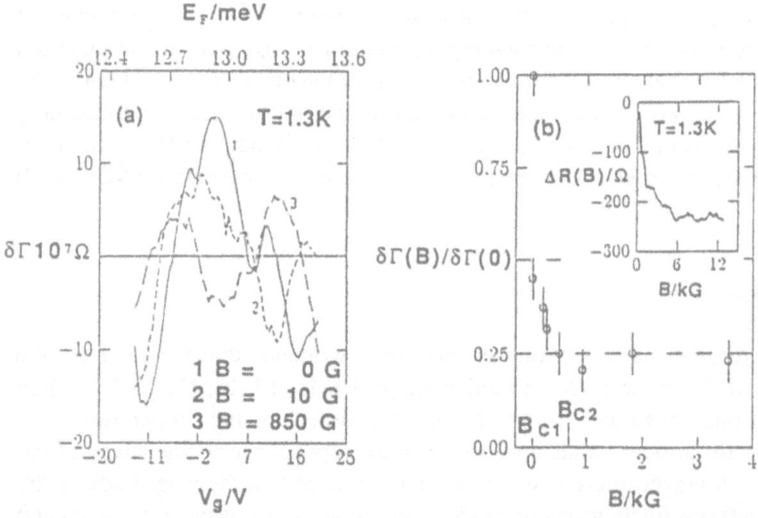

Figure 4 The conductance fluctuations $\delta\Gamma$ of a modulation doped AlGaAs/GaAs heterostructure. (a) $\delta\Gamma$ as a function of the voltage V_g applied to a gate for three different magnetic field strengths. The change of V_g is equivalent to changing the Fermi energy E_F. (b) Variance of $\delta\Gamma(B)$ as a function of the magnetic field. Inset: 'magneto-fingerprint' for $V_g = 10$ V (redrawn after [23]).

2.3 Strong Localization

In insulating systems where transport is governed by hopping processes associated with the coupling of the electronic with the vibrational (or other) degrees of freedom, there are characteristic dependencies on temperature, disorder, magnetic field, and other parameters as, for instance, the density of the charge carriers [26, 27]. In this region quantum coherence is only partly conserved, namely between two successive hopping events. The statistics of the transport can be expected to be a complicated mixture of the statistical properties of the coherent quantum transport and the statistics of the hopping processes. Especially when aiming at quantitative comparisons with experiment, there seems to be at present no generally accepted theoretical model for the understanding of the statistical properties of this regime [28, 29] (Fig.s 5,6).

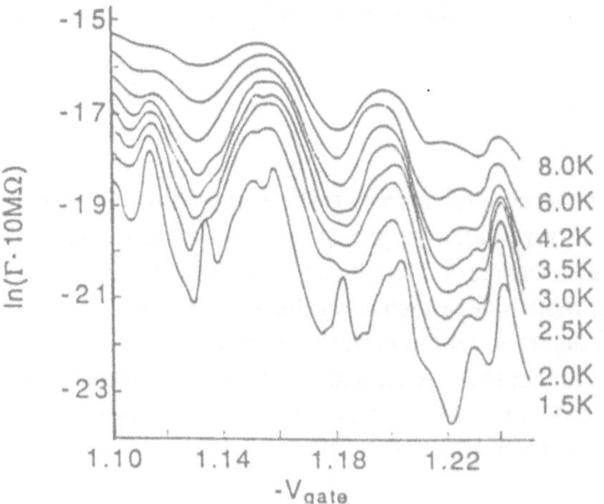

Figure 5
The results for the logarithm of the conductance Γ in units of $10^{-7}\Omega^{-1}$ obtained by Savchenko an his coworkers using a short and wide inversion layer in a GaAlAs/GaAs field effect transistor upon variation of the voltage V_g. Temperatures were as indicated at the curves (redrawn after [26])

2.4 Theoretical Results

Theoretical approaches have included perturbational theories based on quantum interference processes [9, 14, 30, 31, 32, 33], random matrix theoretical treatments using Lyapunov exponents [34, 35] in connection with the principle of maximum entropy [5, ?] for the metallic regime, and field theoretical approaches near the metal-insulator transition [10]. Specific models were investigated in order to achieve a better understanding of the statistical properties in the regime of hopping transport [28, 41, 42, 43]. Direct calculations of the conductance of finite systems by analytical and numerical means were used to understand its quantum statistical properties especially at zero temperature in more detail [29, 44, 45, 46].

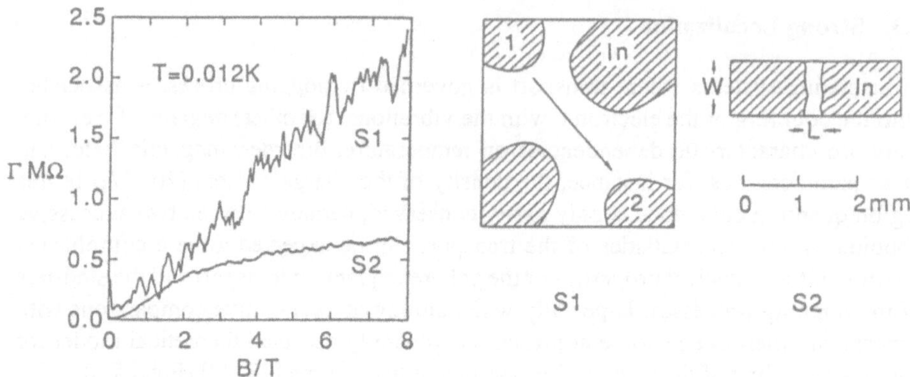

Figure 6 Millikan's results for the magneto-conductance $\Gamma(B)$ (left hand side of the figure) in the hopping regime of two samples of an In_2O_{3-x} film with different geometrical shape (right hand side of the figure) and large transversal geometrical dimensions at a temperature of $T = 0.012$ K (redrawn after [27]).

Metallic Regime

In the metallic limit the moments of the conductance at zero-temperature were calculated by using the diagrammatic, perturbational approach which was also used in the evaluation of the quantum coherence corrections to the conductivity [14, 33]. The 'universality' of the fluctuations, i.e. their independence on the system size, the magnetic field, the temperature, and the disorder, within certain limitations [14, 25], was demonstrated most convincingly by the fact that the second moment was numerically found to be almost independent of the parameters of the system (Fig. 7). Universality was also confirmed by applying random matrix theory [5, 47].

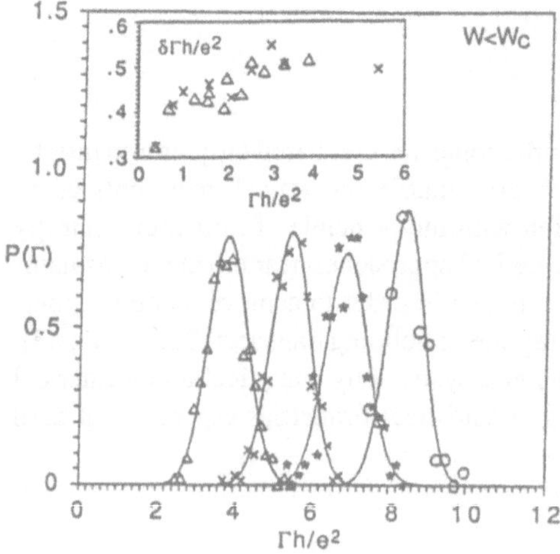

Figure 7
The distribution P of the conductance Γ in units of e^2/h in the metallic regime of a three-dimensional Anderson model for different sizes of the system. The solid lines are Gaussian fits to the data. Insert shows the square root of the variance $\delta\Gamma$ in units of e^2/h for different system sizes (redrawn after [53]).

Insulating Regime

At $T = 0$ the statistics of the conductance of strongly disordered systems (insulating regime) can be characterized by a log-normal distribution function [10, 48, 49] for conductances sufficiently close to, and below its most probable value, mainly from numerical evidence. The logarithm of the conductance is self-averaging. Its relative fluctuations decrease with the size of the system, such that configurational average and most probable value become equal in the thermodynamical limit. The conductance, on the other hand, exhibits extraordinarily strong fluctuations which decrease much slower than the mean value with increasing system size such that the relative fluctuations diverge [28, 29, 45]. As a consequence, configurational average and most probable value do not agree in the thermodynamic limit. Also, configurational averages of the conductance and the resistance are not inverse to each other [50]. Besides the apparent theoretical implications there is the question, which of these is the appropriate quantity to be considered when comparing with experiments [29, 30, 33].

An important and striking observation was that in one dimensional random systems the moments of the conductance divided by its configurational average value tend to be universal [44, 51]. This behaviour was shown to be closely related to the fact that the distribution of the logarithm of the conductance is linear close to zero [29] (Fig. 8).

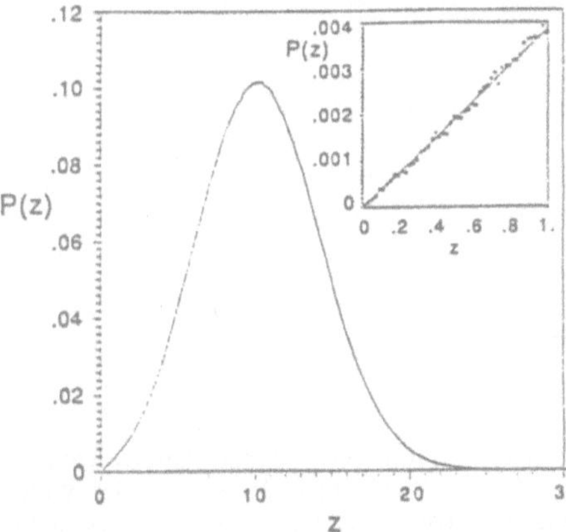

Figure 8
The distribution P of the logarithm of the conductance $z = -\gamma L$ of a one dimensional disordered Anderson model. Inset shows detail for $z \to 0$ (redrawn after [29]).

Critical Point

There is theoretical evidence that the probability distribution of the conductance at the critical point that corresponds to the disorder-induced metal-insulator transition is a universal function, independent of the size of the system, and the position of the critical point at the phase trajectory [52]. The results of a direct numerical calculation obtained for the disordered Anderson model with diagonal disorder that confirms this remarkable suggestion are shown in Fig. 9 [53]. Unfortunately, it is extremely difficult, if not impossible, to verify this directly by an experiment. However, one of the consequences, namely that close to the critical point the logarithm of the conductance, or, more precisely, its most probable value obeys a one-parameter scaling law from which the critical behaviour may be extracted, should be experimentally accessible via measurements of the critical exponent of the dc-conductivity.

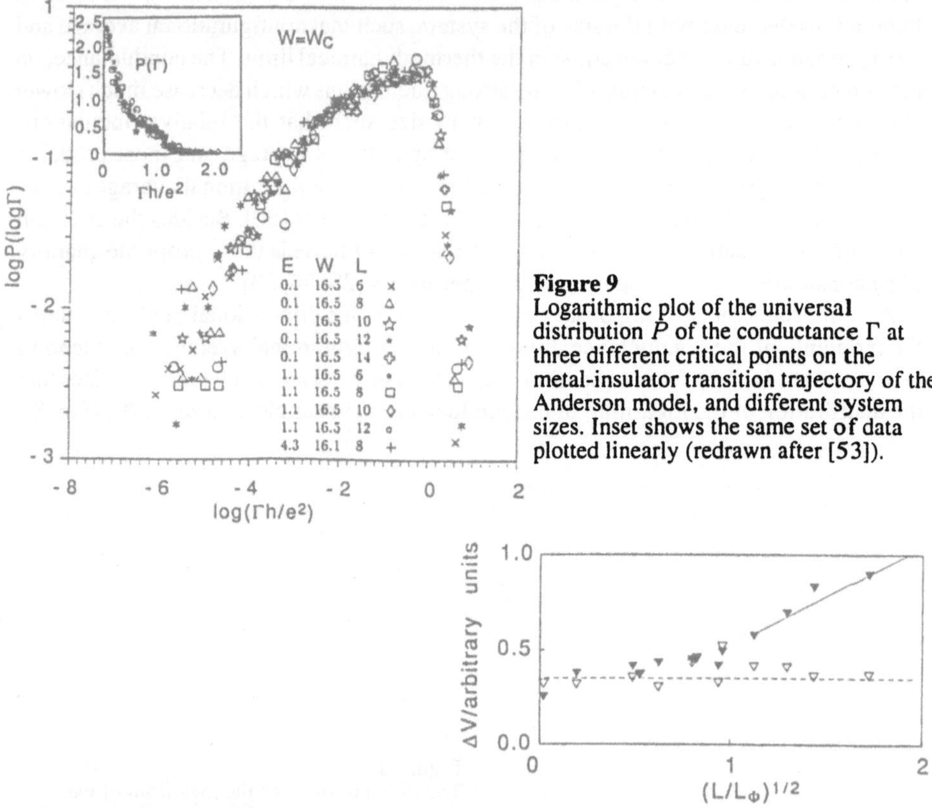

Figure 9 Logarithmic plot of the universal distribution P of the conductance Γ at three different critical points on the metal-insulator transition trajectory of the Anderson model, and different system sizes. Inset shows the same set of data plotted linearly (redrawn after [53]).

Figure 10 Voltage fluctuations ΔV measured in a so-called four-probe experiment by varying an applied magnetic field as a function of $L/L_\Phi(T = 0.04K)$ in a Sb wire of the width 0.1μm. The component symmetric in the magnetic field (black triangles) is almost constant when $L/L_\Phi < 1$, and scales almost classically when $L/L_\Phi > 1$, whereas the antisymmetric component (open triangles) is almost constant. The data were recorded at $T = 0.04K$ (redrawn after [59]).

2.5 Temperature Dependence

In metallic systems the experimental study of the fluctuations at finite temperature T is comparatively easy, provided T is sufficiently low, by varying an external magnetic field or the charge density. The temperature dependent phase coherence length, which is the distance a charge carrier can diffuse coherently between two phase randomizing (inelastic) scattering events [55, 56, 57, 58], can become comparable or even larger than the geometrical size of the system at low temperatures. As a consequence, the conductance fluctuations become insensitive to changes of the temperature (Fig. 10).

The basic assumption that is underlying the comparison between theory and experiment in this regime is that of 'ergodicity' [14]. The fluctuations within a statistical ensemble of macroscopically equivalent systems are equivalent to the fluctuations in a specific member of the ensemble when an external parameter, for instance an applied magnetic field is varied.

In insulators, there is evidence for a strong and significant dependence on the temperature [7, 13, 26, 27]. Since in these systems transport is mediated mainly by quantum mechanically incoherent hopping processes, it is not completely obvious whether or not it is sufficient to take into account only the statistics of the zero-temperature conductance or, equivalently, its logarithm. For the understanding of the experimental data, especially their temperature dependence, a proper treatment of the statistics of the transport processes at finite temperatures seems to be necessary. Also, the above mentioned ergodicity is by no means guaranteed in this case [54].

3 Routes to Qualitative Understanding

Some of the aspects of the reproducible conductance fluctuations can be understood by elementary means. This section serves to outline some basic arguments.

3.1 Classical Limit

In classical statistical physics a physical quantity, say A, is defined in such a way that in the limit of infinite system size L the statistical fluctuations become vanishingly small relative to the average, i. e.

$$\frac{\langle\delta A^2\rangle^{1/2}}{\langle A\rangle} \propto \frac{1}{\Omega^{1/2}}, \qquad \Omega \to \infty, \tag{1}$$

where $\Omega \equiv L^d$ is the volume, $\delta A = A - \langle A\rangle$, and $\langle\cdots\rangle$ denotes the statistical average. A quantity fulfilling (1) is denoted as *self-averaging*. Its average converges towards the most probable value for $\Omega \to \infty$.

The classical conductance Γ is no exception. Since $\langle\Gamma\rangle = \sigma_0 L^{d-2}$ (for a d-dimensional hypercube of diameter L), and $\sigma_0 \propto \tau$, where τ is the mean free time between two successive scattering events, we have $\delta\Gamma \propto \delta\tau$. When we assume that impurities distributed at random with a mean density $n \equiv a_0^{-d}$ are responsible for the scattering,[1] it is plausible that $\tau \propto a_0^d$ (Fig. 11). It follows that

[1] This assumption is justified at low temperatures where inelastic processes are almost completely frozen

$\delta\Gamma/\langle\Gamma\rangle \propto \delta\tau/\tau \propto \delta a_0/a_0 \propto \delta n/n$ [60]. As in an ideal gas the relative fluctuations of the concentration are $\delta n/n \propto L^{-d/2}$, thus, $\delta\Gamma/\langle\Gamma\rangle \propto L^{-d/2}$.

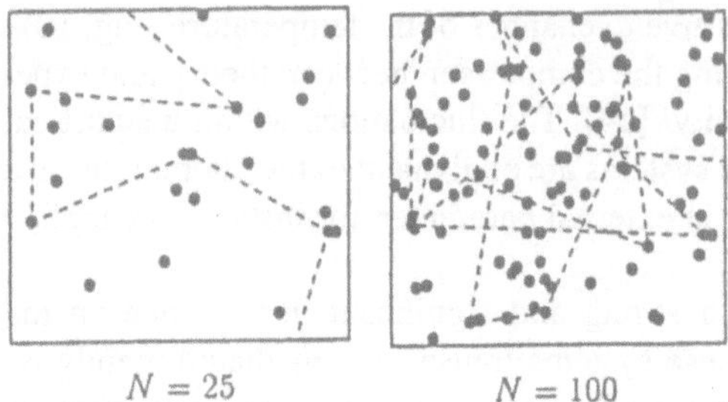

Figure 11
Scattering of a classical particle (dashed trace) at an assembly of point-like scatterers. Low density of scatterers leads to rare scattering events (left) whereas high density causes frequent scatterings (right).

The absolute fluctuations of the conductance decrease with the system size, $\delta\Gamma \propto L^{d/2-2}$. This, if properly evaluated for a specific sample, turns out to be much too small to account for the experimental observations (and theoretical results) in the quantum coherent regime, besides depending on the system size, in contrast to the experimentally observed universality for metals.

3.2 Random Matrix Approach

The universality of the conductance fluctuations in the metallic limit can be understood by using a simple argument borrowed from the theory of random matrices [61]. It gives not only the correct order of magnitude but also sheds some light on their physical origin [47].

The starting point is relation between the quantum conductance of a quasi-1D system of the length L with finite cross-section $\propto N$ (Fig. 12), and its quantum mechanical transmission properties derived by Pichard [34, 35],

$$\Gamma = \frac{2e^2}{h} \sum_{j=1}^{N} \frac{1}{\cosh^2 \gamma_j L}. \tag{2}$$

Here, the factor of 2 is due to spin degeneracy, and $\exp \gamma_j L$ are the eigenvalues of a product of random transfer matrices $Q_L^\dagger Q_L$ with

$$Q_L = \prod_{\nu=1}^{L} T_\nu(E), \tag{3}$$

and

$$T_\nu(E) = \begin{pmatrix} E - H_\nu & -V \\ V & 0 \end{pmatrix}, \tag{4}$$

---

out.

H_ν and V are the disordered Hamiltonian matrix that corresponds to the ν-th slice of the system, and the coupling matrix between the slices, respectively. T_ν couples the amplitudes of the states on the right hand side of the slice to those on the left hand side (Fig. 12).

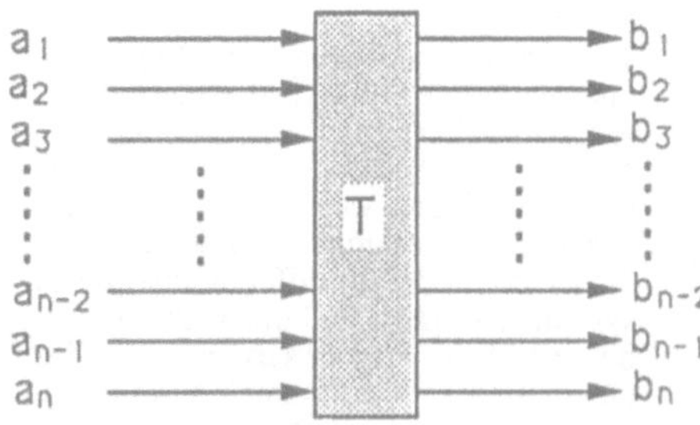

Figure 12
Transmission problem for a quasi-one dimensional system with a finite cross-section. The transfer matrix T represents the system of the cross-section N. It couples the $n = 2N$ amplitudes $b_1 \cdots b_n$ on the right with those, $a_1 \cdots a_n$, on the left hand side.

For very small disorder in H_ν a large but finite number of the Lyapunov exponents γ_j, say N_{eff} will contribute to the conductance (2) each with a contribution of the order $2e^2/h$. They fulfil the condition that $\gamma_j^{-1} > L$. The terms for which $\gamma_j^{-1} < L$ will be exponentially small. Therefore in the metallic regime to a good approximation

$$\Gamma \approx \frac{2e^2}{h} N_{\text{eff}}. \tag{5}$$

If all of the non-vanishing contributions were statistically independent one would expect that the relative fluctuations of the conductance behaved as $N_{\text{eff}}^{-1/2}$ for large N_{eff}. This is in apparent disagreement with experimental, numerical, and other theoretical results. Therefore the γ_j cannot be statistically independent.

The product of the random matrices $Q_L^\dagger Q_L$ is again a random matrix. One may ask whether or not the theory of random matrices designed for matrices with statistically uncorrelated matrix elements is applicable in this situation were the matrix elements are not uncorrelated. In fact, there are strong analytical arguments and numerical evidence [53, 5, 47] that indicate that certain properties of the γ_j can indeed be understood by applying random matrix theory. An important result is that γ_j are in general *not* statistically independent but strongly correlated at least in the metallic limit (Fig. 13).

As a consequence the change in the conductance induced by a small microscopic change in the randomness, e. g. due to the change in the position of a single impurity, cannot be arbitrarily small. Either N_{eff} is unchanged, and the conductance will be essentially the same, or N_{eff} changes by unity such that $\delta\Gamma \approx 2e^2/h$.

We can also understand the magnetic field induced reduction of the fluctuations within this approach. According to random matrix theory the 'level' separation distribution $P(s)$, $s \equiv \gamma_{j+1} - \gamma_j$, behaves as

$$P(s) \propto s^\beta, \tag{6}$$

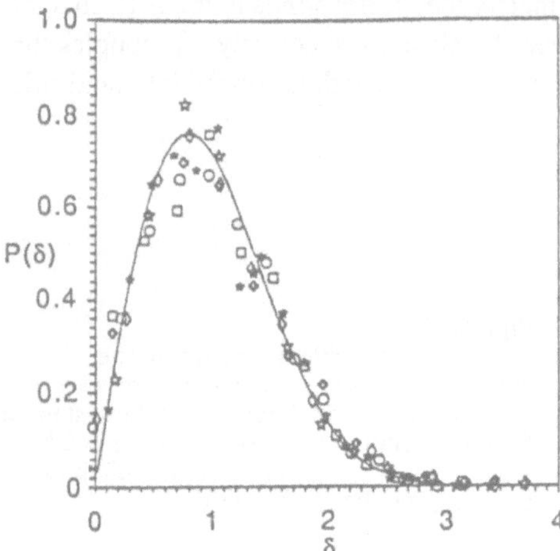

Figure 13
The probability distribution $P(\delta)$ of the distance between the smallest and the second smallest Lyapunov exponent, $\delta = (\gamma_2 - \gamma_1)/\langle\gamma_2 - \gamma_1\rangle$, in the metallic regime. The data were obtained from a quasi-one dimensional disordered Anderson model for different disorders and system sizes [53]. Full line is the result obtained from Wigner surmise [61]. The fact that $P(\delta) \propto \delta$ for small δ indicates that the Lyapunov exponents are repelling each other (redrawn after [53].

for $s \to 0$. Here, $\beta = 1, 2, 4$ depending on whether the system has time-reversal symmetry (orthogonal universality class) or not (unitary universality class), or belongs to the symplectic universality class. The latter will generally be the case in the presence of spin-orbit interaction. Increasing β means physically more 'rigidity' in the spectrum of the eigenvalues of the random matrix, i. e. less randomness (Fig. 14). Thus, the fluctuations of the conductance will generally be reduced when applying a magnetic field to a system belonging to the orthogonal class, since the magnetic field destroys time-reversal invariance and drives the system to unitary symmetry, i. e. $\beta = 2$. Increasing the magnetic field further will eventually lift the spin-degeneracy by Zeeman splitting which results in another decrease of the fluctuations by about a factor of two. The prefactors of the fluctuations can be obtained by by diagrammatic perturbation theory [14]. The random matrix approach, however, provides insight into the deeper origin of the source of the changes in their magnitude with changes in the symmetry properties.

3.3 Dimension One

The paradigm for the statistical behaviour in the strongly localized, insulating region is represented by the conductance of a one-dimensional disordered system. Here, much effort has been devoted to rigorously proof that the 'localization length' λ, which is nothing but the inverse of the (smallest) Lyapunov exponent, is finite for any energy and randomness, when the length of the system tends to infinity [62, 63, 64]. In addition, it was possible to show that the localization length obeys the 'law of large numbers' [65, 66], namely

$$\frac{\delta\lambda}{\langle\lambda\rangle} \propto \frac{1}{L^{1/2}} \qquad (N \to \infty) \tag{7}$$

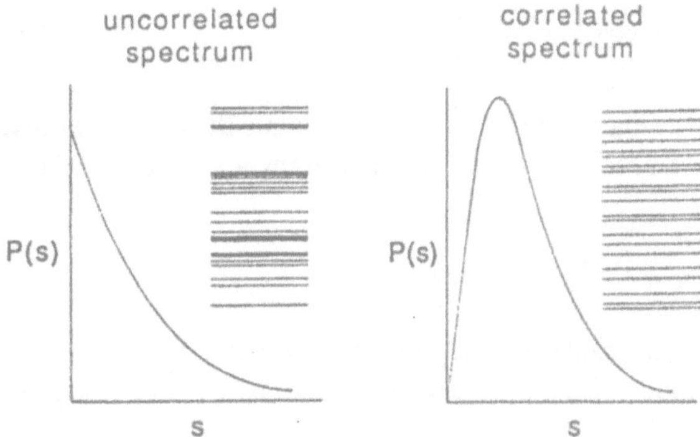

Figure 14
Qualitative picture of the statistical eigenvalue spectra and corresponding level spacing distributions $P(s)$ for orthogonal (left) and unitary symmetry (right).

where $\delta\lambda$ is the square root of the variance of λ.

Applying (2) to one dimension, and noting that the transfer matrices are 2×2 we obtain that for $L \gg \gamma^{-1}$ the conductance is given by [2]

$$\Gamma = \frac{2e^2}{h} e^{-\gamma L}. \tag{8}$$

It is plausible that the conductance as well as the resistance $R \equiv \Gamma^{-1}$ cannot be self-averaging. Small statistical fluctuations in γ will result in exponentially enhanced fluctuations of the conductance or resistance.

In the case of the resistance the proof that this is indeed true is elementary [66]. The results for the first and second moments of the resistance are

$$\langle R(L) \rangle \propto \exp \sigma_2 L/2, \tag{9}$$

and

$$\langle R(L)^2 \rangle \propto \exp \sqrt{3}\sigma_2 L, \tag{10}$$

respectively, for the Anderson model with diagonal disorder of the variance $\sigma_2 (\ll 1)$. The fluctuations $(\langle R^2 \rangle - \langle R \rangle^2)^{1/2}$ of the resistance are increasing exponentially and diverge faster than the configurational average. The latter is therefore not representative of the statistical ensemble, and experimentally of no significance [67].

In the case of the conductance one can show that the moments satisfy [44]

$$\lim_{L \to \infty} \langle \Gamma(L)^n \rangle = C_n \langle \Gamma(L) \rangle. \tag{11}$$

Using the explicit asymptotic form of the distribution of γ which was derived by using numerical methods (Fig. 8) [29],

[2] In general the Lyapunov exponents that correspond to the $2N \times 2N$ transfer matrices come in pairs such that $\gamma_j = -\gamma_{2N-j}$, $(j = 0 \cdots N-1)$. Hence in one dimension which corresponds to $N = 1$ there is only one exponent, $\gamma \equiv \lambda^{-1}$.

$$P_\gamma(\gamma) \propto \gamma \exp\left\{-\frac{(\gamma - \langle\gamma\rangle)^2}{2\Delta^2}\right\}, \tag{12}$$

where $\Delta^2 = 2\langle\gamma\rangle/L$ for sufficiently small disorder, one can calculate the C_n explicitly by applying the method of steepest descent around the maximum $\gamma_m \propto L^{-1}$ of the integrand in $\int d\gamma P_\gamma(\gamma)\Gamma(\gamma L)^n$,

$$C_n \propto n^{-1/2} \cosh^{-2n}(2). \tag{13}$$

We note that the moments of the conductance are completely dominated by extremely small, and thus very improbable values of the Lyapunov exponent in the thermodynamic limit.

Physically this implies that the average of the conductance is dominated by a very few extraordinarily improbable samples with conductances close to $2e^2/h$. They are not representative of the ensemble and practically not accessible in an experiment.

What are then the experimentally observable resistances and conductances?

When preparing a sample the experimentalist 'chooses a specific realization of the randomness' from the statistical ensemble of macroscopically equivalent systems. This sample will yield with a very high probability a conductance or resistance close to the *most probable value* (Fig. 15). The most probable values correspond approximately to exp $\langle\Gamma\rangle$. It is therefore the logarithm of the conductance or the resistance which is of physical relevance, and not the configurational average [67].

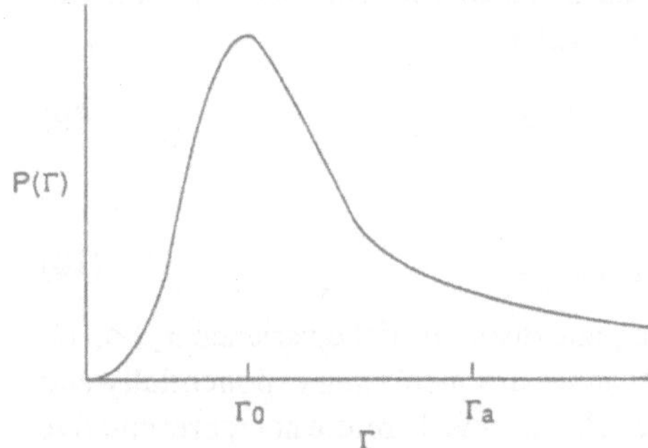

Figure 15
Probability distribution of the conductance of a disordered one dimensional system, schematically. The most probable value, $\Gamma_0 \approx \exp\langle\ln\Gamma\rangle$, is experimentally accessible, and not the configurational average $\Gamma_a \equiv \langle\Gamma\rangle$.

3.4 'Ergodic Principle'

In the previous sections we have discussed the fluctuations within a statistical ensemble of macroscopically equivalent systems. How can these be related with the experimentally observed conductance fluctuations when an external parameter is varied for a given sample? It is not immediately obvious why varying the strength of an applied magnetic field, for instance, should be equivalent to changing the randomness, which is equivalent to choosing a different member of the ensemble.

For the fluctuations induced by changing the charge density, as in the above mentioned MOSFET experiment, it is at least plausible that variations in the charge density are equivalent to variations of the randomness. Changing the charge density is equivalent to changing the Fermi energy where transport takes place at low temperature. The charge carriers participating in the transport processes will therefore 'see' a different randomness when changing the Fermi energy like a hiker experiences a different shoreline in a rocky landscape filled with water when the water level is varied.

This leads to an 'ergodic' hypothesis in connection with the transport fluctuations. It states that the ensemble average of any statistical property of the conductance as, for instance, the second moment or a two-point correlation function, is equivalent to the average over a sufficiently large range of variation of the external parameter(s).

The ergodic hypothesis is very plausible for variations of the charge density or Fermi energy. There is also theoretical evidence for its approximate validity [14, 30]. When varying the magnetic field the connection is more difficult to establish, though. To some extent, when the charge density is fixed, a variation in the strength of an external magnetic field will lead to changes in the random energy spectrum in such a way that the Fermi energy will be shifted against the stochastic potential induced by the impurities, and the above argument can again be applied. That the situation is more difficult than suggested by this simple argument is obvious from the changes in the magnitude of the fluctuations induced by a magnetic field described in the previous section. It has been also explicitly demonstrated experimentally by investigating the conductance fluctuations of an inversion layer in a short and wide GaAs heterostructure upon varying an external magnetic field, and, alternatively, a gate voltage which served to change the charge density in the inversion layer [54].

3.5 Phase Coherence

The conductance fluctuation phenomena are observed at very low temperatures. The reason is that here inelastic processes are sufficiently suppressed not to disturb the quantum mechanical motion of the electrons within the sample, as mentioned above.

This argument can be made more quantitative. Assume that the electron's motion in the random impurity potential is equivalent to a random walk with a certain mean free path $\ell = v_F \tau$ where τ is the mean time between two collisions with impurities. The distance the electron can diffuse elastically without suffering an inelastic, 'phase-randomizing' process is then given by $L_\Phi = \sqrt{D_0 \tau_\Phi}$ where $D_0 (\neq 0)$ is the zero-temperature diffusion constant related to the dc-conductivity by Einstein's relation, and τ_Φ the mean time between inelastic or, more precisely, phase breaking collisions. L_Φ increases with decreasing temperature since inelastic scattering events become suppressed at lower temperatures. When L becomes of the order of or larger than the geometrical size of the system L, the reproducible fluctuations will be observable, and will be essentially temperature independent. Another condition for the observability of the fluctuations is that $L \approx L_\Phi \gg \ell$.

When the impurity scattering is sufficiently strong the system undergoes a metal-insulator transition. Then the localization length $\lambda(> \ell)$ is much smaller than L. No diffusion takes place in this regime, $D_0 = 0$, transport will be governed by hopping involving inelastic processes as a necessary ingredient. Presently, it is not completely clear how to obtain the temperature dependence of the experimentally observed fluctuations in this regime although a number of first attempts were already made [68, 69]. The assumption that the phase coherence length L_Φ is given by the Mott hopping distance leads to results that are not inconsistent with the available experimental findings [28, 29]. However, more experimental data are needed, in order to make quantitative comparisons possible.

4 Summary and Conclusion

In this article I have sketched the present status of the physics of the reproducible quantum fluctuations of the conductance of disordered electron systems with geometrical dimensions of the order of a few micron. They are observable at temperatures close to the absolute zero where inelastic processes that destroy the phase coherence of the quantum states at higher temperatures are suppressed. The fluctuations show qualitatively different behaviours in the metallic and the insulating regimes.

They are 'universal', i. e. quite insensitive to changes in the geometrical dimensions, the microscopic disorder, and the temperature in the metallic limit. Here, a simple route to a qualitative understanding is provided by random matrix theory which explains the order of magnitude by the rigidity of the spectrum of the Lyapunov exponents that characterize the quantum mechanical transmission properties of the system. It explains also the changes in their magnitude that are observed experimentally when varying an applied magnetic field. An 'ergodic' hypothesis connects the fluctuations observed experimentally on a given sample when an external system parameter is varied with the sample-to-sample fluctuations in the statistical ensemble. The condition for their observability in this regime is that the temperature dependent phase breaking length is of the order of or larger than the geometrical size of the system. The theoretical understanding of the universal conductance fluctuations seems presently to be rather complete, although a commonly accepted approach for the phase breaking length is still lacking, and the 'ergodic hypothesis' is only qualitatively understood.

On the other hand, in the insulating regime the relative fluctuations of the quantum conductance diverge in the thermodynamic limit. The behaviour of the conductance is therefore totally determined by the fluctuations. As a consequence, configurational averages cannot be assumed to be representative for the transport behaviour observed in experiments done on specific samples at very low temperatures. Instead, the most probable value of the conductance has to be used as a starting point. Most surprisingly, reproducible fluctuations of the magneto-conductance were observed in thin insulating

In_2O_{3-x}-films with sizes of the order of a *few millimeters*. They depend in a characteristic way on the temperature, and are only one example for experimental findings in the regime of hopping transport that are far from being theoretically well understood. In this regime the concept of the phase breaking length breaks presumably down. In addition, there is experimental evidence that indicates that the 'ergodic hypothesis' might not be valid.

Reproducible conductance fluctuations, although looking at the first glance like an experimental artifact, provide thus a useful tool for a deeper understanding of the quantum transport properties of random systems. In addition, they allow for a systematic study of the properties of physical quantities that do not obey the conventional rules of Statistical Physics. Their thorough experimental and theoretical investigation is likely to contribute considerably towards our insight into the microscopic electronic processes in disordered matter.

Acknowledgement
I wish to thank Dirk Endesfelder for critically reading the manuscript.

Bibliography

[1] D. S. Betts *An Introduction to Millikelvin Technology* (Cambridge University Press, Cambridge 1965)

[2] H. Heinrichs, G. Bauer, F. Kuchar (eds) *Physics and Technology of Submicron Structures* Springer Ser. Sol. St. Sci. **85** (Springer Verlag, Berlin 1988)

[3] G. Bergmann, Phys. Rep. **107** 1 (1984)

[4] D. Vollhardt, P. Wölfle, in: *Electronic Phase Transitions* W. Hanke, Yu. V. Kopaev (eds) p. 1 (Elsevier Science Publishers, 1992)

[5] J.-L. Pichard, in *Quantum Coherence in Mesoscopic Systems* edited by B. Kramer, NATO ASI Ser. B **254** 369 (1991)

[6] A. MacKinnon, J. Phys. **C13** L1031 (1980)

[7] A. B. Fowler, A. Hartstein, R. A. Webb, Phys. Rev. Lett. **48** 196 (1982)

[8] D. Mailly, M. Sanquer, in *Quantum Coherence in Mesoscopic Systems* edited by B. Kramer, NATO ASI Ser. B **254** 401 (1991)

[9] A. G. Aronov, Yu. V. Spivak, Rev. Mod. Phys. **59** 755 (1987)

[10] I. V. Lerner, in: *Quantum Coherence in Mesoscopic Systems* edited by B. Kramer, NATO ASI Ser. B **254** 279 (1991)

[11] R. Blümel, U. Smilansky, Phys. Rev. Letters **64** 241 (1990); U. Smilansky, Lecture Notes of Miniworkshop on Quantum Chaos ICTP Trieste (1990)

[12] T. Geisel, R. Ketzmerick, G. Petschel, Phys. Rev. Letters **66** 1651 (1991)

[13] S. Washburn, R. A. Webb, Adv. Phys. **35** 375 (1986)

[14] P. A. Lee, A. D. Stone, H. Fukuyama, Phys. Rev. B **35** 1039 (1987)

[15] D. Popović, A. B. Fowler, S., Washburn, Phys. Rev. Letters **67** 2870 (1991)

[16] R, G, Wheeler, K. K. Choi, A. Goel, R. Wisnieff, D. E. Prober, Phys. Rev. Letters **49** 1574 (1982)

[17] C. P. Umbach, S. Washburn, R. B. Laibowitz, R. A. Webb, Phys. Rev. B**30** 4048 (1994)

[18] W. J. Skocpol, L. D. Jackel, R. E. Howard, H. G. Craighead, L. A. Fetter, P. M. Mankiewich, P. Grabbe, D. M. Tennant, Surf. Sci. **142** 14 (1984)

[19] W. J. Skocpol, P. M. Mankiewich, R. E. Howard, L. D. Jackel, D. M. Tennant, A. D. Stone, Phys. Rev. Letters **56** 2865 (1986)

[20] J. C. Licini, D. J. Bishop, M. A. Kastner, J. Melngailis, Phys. Rev. Letters **55** 2987 (1985)

[21] J. Caro, J. A. Gao, A. H. Verbruggen, S. Radelaar, J. Middelhoek, in *Quantum Coherence in Mesoscopic Systems* edited by B. Kramer, NATO ASI Ser. B **254** 405 (1991); J. R. Gao, A. H. Verbruggen, S. Radelaar, J. Middelhoek, Phys. Rev. B**40** 11676 (1989-I)

[22] D. Mailly, M. Sanquer. J. L. Pichard, P. Pari, Europhys. Lett. **8** 471 (1989)

[23] P. Debray, J. L. Pichard, J. Vicente, P. N. Tung, Phys. Rev. Letters **63** 2264 (1989)

[24] J. L. Pichard, M. Sanquer, K. Slevin,, P. Debray, Phys. Rev. Letters **65** 1812 (1990)

[25] S. Iida, H. A. Weidenmüller, J. A. Zuk, Phys. Rev. Letters **64** 583 (1990); Ann. Phys. **200** 219 (1990)

[26] A. O. Orlov, M. E. Raikh, I. M. Ruzin, A. K. Savchenko, Sov. Phys. JETP **69** 1229 (1989)

[27] M. L. Milliken, Z. Ovadyahu, Phys. Rev. Lett. **65** 911 (1990)

[28] B. Kramer, A. Kawabata, M. Schreiber, Phil. Mag. B**65**, 595 (1992)

[29] P. Markos, B. Kramer, Annalen der Physik **2** 339 (1993)

[30] P. A. Lee, A.D. Stone, Phys. Rev. Lett. **55** 1622 (1985)

[31] A. D. Stone, P.A. Lee, Phys. Rev. Lett. **54**, 1196 (1985)

[32] B. L. Altshuler, D. E. Khmelnitskii, JETP Lett. **42** 291 (1985)

[33] B. L. Altshuler, V. E. Kravtsov, I. V. Lerner, Sov. Phys. JETP **64** 1352 (1986); Phys. Letters A**134** 488 (1989)

[34] J.-L. Pichard, PHD Thesis, University of Paris Orsay n^0 2858 (1984)

[35] J.-L. Pichard, G. André, Europhys. Lett. **2** 477 (1986)

[36] J. L. Pichard, N. Zanon, Y. Imry, A. D. Stone, J. Phys.(Paris) **51** 587 (1990)

[37] P. A. Mello, Phys. Rev. B**35** 1082 (1987-II); Phys. Rev. Letters **60** 1089 (1988)

[38] P. A. Mello, E. Akkermans, B. Shapiro, Phys. Rev. Letters **61** 459 (1988)

[39] P. A. Mello, P. Pereyra, N. Kumar, Ann. Phys. **181** 290 (1988)

[40] P. A. Mello, A. D. Stone, Phys. Rev. B**44** 3559 (1991-II)

[41] V. L. Nguyen, B. Z. Spivak, B. I. Shklovskii, JETP Lett. **43**, 44 (1986); Sov. Phys. JETP **62**, 1021 (1986)

[42] R. A. Serota, R. K. Kalia, P. A. Lee, Phys. Rev. B **33** 8441 (1986)

[43] M. E. Raikh, I. M. Ruzin, Sov. Phys. JETP **65** 1273 (1987); **68** 642 (1989); and in: *Mesoscopic Phenomena in Solids*, B. L. Altshuler, P. A. Lee, R. A. Webb (eds.), (North Holland 1991)

[44] J. B. Pendry, A. MacKinnon, P. J. Roberts, Proc. Roy. Soc. (London) **537**, 67 (1992)

[45] K. S. Chase, A. MacKinnon, J. Phys. C: Solid State Physics **20** 6189 (1987)

[46] E. Medina, M. Kardar, Phys. Rev. B**46** 9984 (1992-II)

[47] Y. Imry, Europhys. Lett. **1** 249 (1986)

[48] B. Shapiro, Phys. Rev. Lett. **65** 1510 (1990)

[49] A. Cohen, Y. Roth, B. Shapiro, Phys. Rev. B **38**, 12125 (1988)

[50] J. Sak, B. Kramer, Phys. Rev. B**24** 1761 (1981)

[51] N. Giordano, Phys. Rev. B**38** 4746 (1988)

[52] B. Shapiro, Phil. Mag. B**56** 1032 (1987)

[53] P. Markoš, B. Kramer, Phil. Mag. B**68** 357 (1993)

[54] A. O. Orlov, A. K. Savchenko, A. V. Koslov, Sol. St. Commun. **78** 743 (1989)

[55] D. J. Thouless, Phys. Rev. Letters **39** 1167 (1977)

[56] P. W. Anderson, E. Abrahams, T. V. Ramakrishnan, Phys. Rev. Letters **43** 718 (1979)

[57] D. J. Thouless, Sol. St. Commun. **34** 683 (1980)

[58] B. L. Altshuler, A. G. Aronov, in: Electron-Electron Interactions in Disordered Systems, ed. by A. L. Efros and M. Pollak, 1 (North Holland 1985)

[59] S. Washburn, IBM J. Res. Dev. **32** 335 (1988)

[60] D. A. Stone, in reference [2] p. 108

[61] M. L. Mehta, *Random Matrices and the Statistical Theory of Energy Levels* (Academic Press 1967)

[62] N. F. Mott, W. D. Twose, Adv. Phys. **10** 107 (1961)

[63] K. Ishii, Progr. Theor. Phys. (Suppl.) **53** 77 (1973)

[64] P. Erdös, R. C. Herndon, Adv. Phys. **31** 65 (1982)

[65] A. J. O'Connor, Commun. Math. Phys. **45** 63 (1975)

[66] E. Abrahams, M. Stephen, J. Phys. C**13** L377 (1980)

[67] P. W. Anderson, D. J. Thouless, E. Abrahams, D. S. Fisher, Phys. Rev. B **22**, 3519 (1980)

[68] J.-L. Pichard, Inst. Phys. Conf. Ser. **108** 77 (1990)

[69] S. Feng, J.-L. Pichard, Phys. Rev. Lett. **63** 753 (1991)

[45] K. S. Chase, A. MacKinnon, J. Phys. C: Solid State Physics 20, [illegible] (1987)

[46] [illegible], M. Kramer, Phys. Rev. B46 9944 (1992) [illegible]

[47] Y. Imry, Europhys. Lett. 1 249 (1986)

[48] B. Shapiro, Phys. Rev. Lett. 65 1510 (1990)

[49] A. Cohen, Y. Roth, B. Shapiro, Phys. Rev. B 38 12125 (1988)

[50] J. Sak, B. Kramer, Phys. Rev. B24 1761 (1981)

[51] N. Giordano, Phys. Rev. B38 4746 (1988)

[52] [illegible], Phil. Mag. B50 [illegible] (1984)

[53] P. Markos, B. Kramer, Phil. Mag. B68 357 (1993)

[54] [illegible]

[55] D. J. Thouless, Phys. Rev. Lett. 39 1167 (1977)

[56] [illegible]

[57] [illegible]

[58] B. L. Altshuler, [illegible] Electron-Electron Interactions in Disordered Systems [illegible] North Holland (1985)

[59] [illegible] 32 [illegible] (1968)

[60] [illegible]

[61] M. L. Mehta, Random Matrices and the Statistical Theory of Energy Levels (Academic Press 1967)

[62] N. F. Mott, [illegible] Adv. Phys. 16 49 (1967)

[63] [illegible]

[64] [illegible]

[65] [illegible]

[66] [illegible]

[67] B. W. [illegible], Phys. Rev. B 47 [illegible] (1993)

[68] [illegible] (1993)

[69] [illegible], Phys. Rev. Lett. [illegible]

Raman Scattering in II-VI Compounds

G. Schaack

Institut fü Experimentelle Physik
Am Hubland, D 97074 Würzburg, Germany

1 Introduction

The interest in binary semiconductor compounds comprising elements of the second and the sixth column of the periodic system has a long lasting history [1], however recent developments, culminating in the feasability study of a blue laser diode [2] with its considerable technological impact, lead to a larger concentration of instrumental and manpower investments than ever met before in this field. In such a period of rapid technological developments, all kinds of characterization methods for materials (bulk, epitaxial layers, diode structures, and heterostructures) clearly are of interest and are used for the optimization of growth conditions and, more generally, for the study of material parameters, of elementary excitations, and of important interaction mechanisms.

The huge amount of information and physical insight compiled over the years for the III-V series, especially for GaAs usually cannot be transferred without serious reconsideration to the II-VI compounds. The physical reasons for this difference are mainly due to the nature of the chemical bonds in the II-VI group, which tend to be more ionic than in the III-V compounds. Related with this fact is the increased tendency in the II-VI to form lattice defects of various kinds, especially at interfaces. Consequently, characteristic differences in the band structure exist, of special interest for applications is the tunability of the fundamental band gap in some ternary II-VI compounds.

Raman spectroscopy, especially resonance Raman spectroscopy, is a very efficient tool for material characterization. It probes several important properties of a semiconductor: the energies of low lying excitations (single-particle or collective, e.g. phonons, electronic and magnetic excitations (plasmons, polarons, polaritons, magnons in semimagnetic materials), inter(sub-)band transitions, electronic spin flips, etc. It also delivers information on the type and on the energies of the intermediate excited states involved in the scattering process (e.g. free and bound exciton states, band edge states and donor states) and on interaction processes between the charge carriers and the excitations under study. These interaction processes are of fundamental importance in semiconductor physics: exchange, dipolar and contact interactions, various electron-phonon interaction processes, e.g. deformation potential and Fröhlich interaction.

The full potential of Raman spectroscopy can only be exploited, when this technique is combined with other spectroscopic methods, e.g. luminescence spectroscopy and photo-luminescence excitation spectroscopy (PLE), where the luminescence is detected at a fixed photon energy, while the excitation by a tunable laser is scanned over an interval of shorter wavelengths. It is a great advantage of Raman scattering techniques, that this type of spectroscopy with its elaborate spectroscopic and very sensitive detection equipment can easily perform luminescence and PLE experiments, often with the same sample mount, while the usual instrumentation for the detection of luminescence is unable to detect the Raman signals at low light levels.

Some excellent review articles on the general aspects of resonance phenomena in Raman scattering in semiconductors have been published in the past [3,4]. Based on these references, we can afford to compile here only some basic relations of general relevance for the following discussions.

The important subgroup of (diluted) semimagnetic II-VI semiconductors form ternary compounds $A_{1-x}D_xC$, where a sizable part x of the A_{II} ions in the $A^{II}C^{VI}$ host material is replaced by paramagnetic ions D of the transition metal series on regular lattice sites, preferentially by Mn^{2+}, but Fe^{2+} is also of importance. These materials have attracted the attention of a more fundamentally oriented research community, due to their very broad spectrum of basically interesting magnetic phenomena: various kinds of magnetic order are encountered on changing either the concentration x at fixed temperature T or varying T at constant x (paramagnetic and spin-glass phases, antiferromagnetic ordering, magnetic polaron etc.) [5]. The large x- and magnetic field (B) dependence of the fundamental band gap at Γ is also of interest for various applications. Again (resonance) Raman spectroscopy in these semimagnetic II-VI compounds plays an important role and has been reviewed repeatedly in an excellent and exhausting manner by Ramdas and Rodriguez [6,7].

This article is intended to review some activities of current interest in this field. The author hopes to achieve this task by concentrating on various aspects of the work done at present in his own group, but references to parallel investigations of other groups are also given. After compiling some basic relations of the microscopic theory of Raman scattering for the off-resonance and the in-resonance case, we start with a report on the determination of the absolute Raman scattering cross section of LO phonons in HgTe in the energy region of the E_1 and the $E_1 + \Delta_1$ band gaps and a discussion of the various electron- phonon interaction mechanisms contributing to the scattered amplitude in this case. Numerical values for the deformation potential in this compound will be given. Next we concentrate on a detailed investigation of the electronic spin flip in paramagnetic (Cd,Mn) Te and on the optically detected paramagnetic resonance signal (PRS) from the localized $3d^5$-electrons of Mn^{2+} in the same material. Finally some results on the changes of the LO phonon spectrum in heavily n-doped ZnSe epilayers are presented.

2 Fundamentals of First-Order Resonance Raman Scattering in crystals

Light is inelastically scattered by an optical phonon or some other quasiparticle via the excitation of an electron-hole pair due to electron-photon interaction H_{eR} bringing the crystal to the virtually excited (intermediate) electronic state $|\alpha\rangle$. The scattering of either the electron in state $|\alpha\rangle$ or the hole by the creation or annihilation of the excitation investigated (e.g. via the electron-phonon interaction H_{eL}) transfers the system into the excited state $|\beta\rangle$. The emission of the scattered photon is due to the recombination of this electron-hole pair from state $|\beta\rangle$ back to the electronic ground state $|0\rangle$, (1). Energy is conserved only for the total process, in the case of creation (annihilation) of the excitation with energy $\hbar\Omega$ (Stokes (Anti-Stokes) scattering) the energy difference of the incident ($\hbar\Omega_L$) and scattered photon ($\hbar\Omega_S$) equals +(-) $\hbar\Omega$. The interaction process H_{eL} may involve different mechanisms, which are of basic interest in the following and which can in many cases be discriminated against by different polarization properties of the incoming (L) and outgoing radiation (S) due to the symmetry properties of the Raman tensor $\boldsymbol{R}^{(1)}$, which is a symmetric second rank tensor with six independent components, some of which may be zero for symmetry reasons. In the case of magnetic dipole scattering (e.g. from magnetic excitations) however, the Raman tensor is an antisymmetric traceless tensor with three components (axial vector). The polarization ($\boldsymbol{e}_L$, $\boldsymbol{e}_S$) and orientation of the wavevectors $\boldsymbol{k}_L$, $\boldsymbol{k}_S$ of the optical radiation with respect to the crystal axis will be given in the usual Porto notation in the following: $\boldsymbol{k}_L(\boldsymbol{e}_L, \boldsymbol{e}_S)\boldsymbol{k}_S$.

The scattering cross section $(dS/d\Omega)^{(1)}$ of the first-order Raman process is derived in third order perturbation:

$$\left(\frac{dS}{d\Omega}\right)^{(1)} = \frac{\omega_S^3\omega_L}{c^4}\,\frac{\hbar}{2v_cM^*\Omega}\,\frac{n_S}{n_L}\left|\hat{e}_S\cdot\boldsymbol{R}^{(1)}\cdot\hat{e}_L\right|^2[n(\Omega)+1], \tag{1}$$

$$\left|\hat{e}_S\cdot\boldsymbol{R}^{(1)}\hat{e}_L\right|^2 = \frac{n_L^2n_S^2}{\omega_L^2}\,\frac{2v_cM^*\Omega}{[n(\Omega)+1]}\,\frac{V}{\hbar^3(2\pi)^2}\left|W_{fi}^{(1)}(\hat{e}_S,\hat{e}_L)\right|^2, \tag{2}$$

$$W_{fi}^{(1)} = \sqrt{n(\Omega)+1}\sum_{\alpha,\beta}\frac{\langle 0|H_{eR}^{+}|\beta\rangle\langle\beta|H_{eL}|\alpha\rangle\langle\alpha|H_{eR}^{-}|0\rangle}{(E_\beta-\hbar\omega_S)(E_\alpha-\hbar\omega_L)}. \tag{3}$$

Here v_c is the volume of the primitive unit cell and V is the sample volume, M^* is the reduced mass of this cell; $n(\Omega)$ is the phonon occupation number. n_S, n_L are refractive indices. The transition probability $W_{fi}^{(1)}$ has to be amended by five additional terms derived by various permutations of the sequence of H_{eR} and H_{eL} and by considering the pertinent denominators. The term shown in (3), however, provides the largest contribution and is the only one of interest in the case of resonance. In this case the summation over the intermediate states (α, β) is reduced to the states in or close to resonance.

Resonance occurs, if one of the photon energies $\hbar\Omega_L$ or $\hbar\Omega_S$ coincides with the energy differences E_α, E_β between the intermediate states $\langle\alpha|$ or $\langle\beta|$ of the crystal and the initial or final states $|i\rangle$ or $|f\rangle$. The Raman cross-section is maximum for incoming resonance: $E_\alpha = \hbar\Omega_L$, or outgoing resonance: $E_\beta = \hbar\Omega_S$. In this case the widths of the intermediate states have to be considered by substituting $E_{\alpha,\beta}$ by $E_{\alpha,\beta} - i\Gamma_{\alpha,\beta}$. Fig. 1 gives an example of the resonantly enhanced 1LO and 2LO phonon scattering near the E_1-gap of HgTe. In the special case, where $\hbar\Omega \approx E_\alpha - E_b$, a double resonance may occur with an especially pronounced anomaly of the intensity.

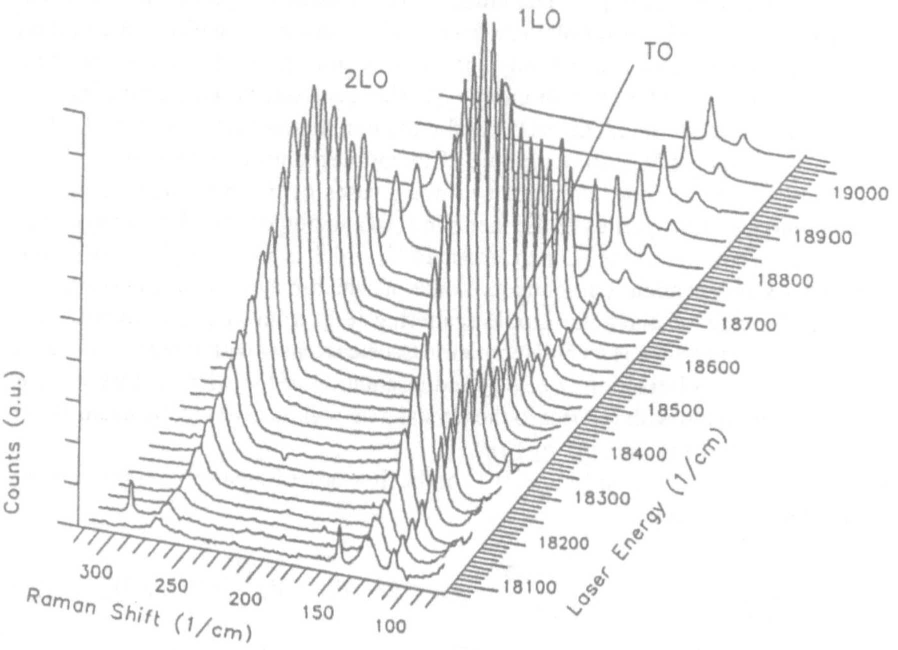

Figure 1 Resonantly enhanced Raman scattered intensity of a [111]-oriented HgTe epilayer near the E_1 gap (18 107 cm^{-1}) in backscattering geometry $-z(x,x)z$ at $T = 1.5$ K as a function of the quantum energy of the incident laser. 1TO phonon at 116 cm^{-1}, 1LO phonon at 136 cm^{-1}, 2LO phonon at 274 cm^{-1}.

3 Resonant LO and TO Phonon Raman Scattering near the E_1 and $E_1 + \Delta_1$ Gaps in HgTe

In most experimental situations the study of relative Raman intensities is sufficient to derive the required information. Such data give the energies and the widths of the excitations under study and, when polarized light is used in the scattering experiment, their types of symmetry. More detailed information on the scattering process and the underlying electron excitation interaction process can be obtained by a measurement of the absolute sizes of the Raman scattering cross sections as a function of the wavelength and by a fit of the pertinent interaction constants occurring in the theories of the scattering process to the experimental results. Using the detailed theoretical expressions available e.g. for the phonon scattering processes since about 10 years, the deformation potential interaction constants, the strength of the Fröhlich contribution and the electro-optic part have been determined for several semiconducting materials recently [see e.g.: 8,9,10]: Here a report is given on a detailed study of the absolute scattering cross sections in HgTe. This substance is different from those previously studied, as it is a zero-gap material with an inverted band structure at Γ, a large dielectric constant, small effective masses and excitonic effects play a minor role. This fact supports the use of uncorrelated electron-hole pairs as the intermediate states in the calculations, a fundamental approximation generally made in the theoretical derivation of the scattering cross section by perturbation calculations. In HgTe the E_1 and the $E_1 + \Delta_1$ gaps along the L line near the L point of the B.Z. lie in the energy range of easily tunable dye lasers.

Several scattering mechanisms resonant near the E_1 and the $E_1 + \Delta_1$ gaps must be taken into account for the LO and TO phonons. There are two-band and three-band contributions to the deformation potential dipole-allowed scattering by TO and LO phonons and the standard intraband dipole-forbidden scattering (LO). The Fröhlich interband or electro-optic mechanism is usually assumed to be a small correction to the deformation potential scattering and gives rise to the differences in scattering efficiencies between LO and TO phonons. Other scattering mechanisms are induced by surface electric fields and by impurities (dipole-forbidden). Only the latter mechanism of both will be of some importance in the present example.

The Raman measurements were performed in backscattering geometry at 1.8 K, and the sample-substitution method was used to eliminate the wavelength dependent characteristics of the experimental set-up in order to obtain absolute values for the Raman tensor. In this method the scattered intensity of the sample is compared with the intensity from a standard with known Raman efficiency under identical experimental conditions. The count rate of the probe is found to be:

$$I_{Pr} = D(\omega_S)\frac{dS}{d\Omega}P_L(0,\omega_L)K(\omega_L,\omega_S)\Omega'. \tag{4}$$

Here $dS/d\Omega$ is the scattering efficiency of the sample, $P_L(0,\omega_L)$ is the intensity of the incident laser beam, $K(\omega_L,\omega_S)$ is a correction function taking the reflectivity and the absorptivity of the sample into account, Ω' is the spherical angle in the sample collected by the spectrometer, and $D(\omega_S)$ describes the unknown characteristics of the spectrometer and other components of the set-up which will be used under identical conditions for the measurements with the standard. The LO phonon signal from a single crystal silicon surface in (001)-direction has been used for this purpose. The following expression for the unknown scattering efficiency will be obtained

$$\frac{dS}{d\Omega}(\omega_L) = \frac{K(\omega_L,\omega_S)_{Standard}}{K(\omega_L,\omega_S)} \frac{dS}{d\Omega}(\omega_L)_{Standard} \frac{I(\omega_L)}{I(\Omega_L)_{Standard}}. \tag{5}$$

For satisfying results, the function K of the probe must be reliably known. The optical constants entering K can be determined from a measurement of the reflectivity of the sample near the E_1 and the $E_1 + \Delta_1$ gap and assuming certain relations holding for the complex electronic susceptibilities of the material [4], which depend on the type of critical points of the density of states at the E_1 and the $E_1 + \Delta_1$ gaps. In Fig. 2 the observed reflectivity at normal incidence of HgTe(111) and the real and imaginary parts of the dielectric function ϵ derived are shown.

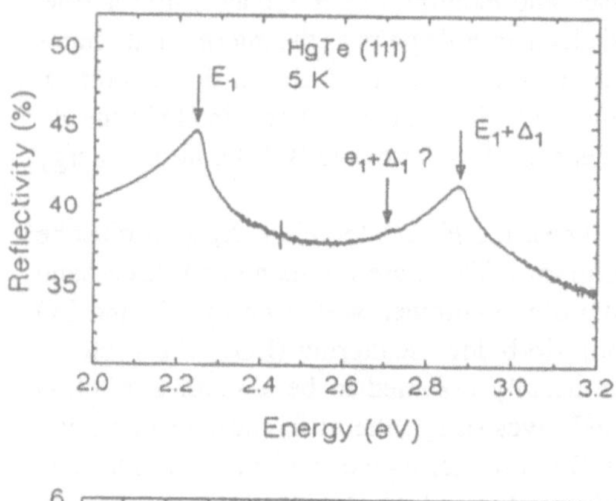

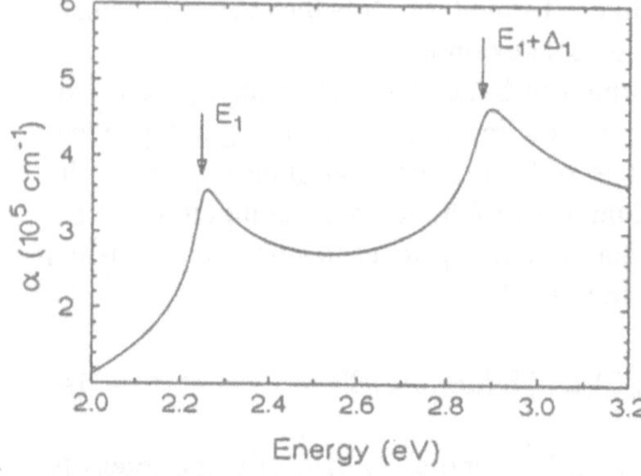

Figure 2
Reflectivity of HgTe (111)-face in the region of the E_1 and $E_1 + \Delta_1$ gaps (above) and the absorption constant α derived.

We now turn to a discussion of the different scattering mechanisms for optical phonons effective in HgTe. The electron-phonon interaction in this polar material is mainly due to the short range deformation potential interaction and the long range Fröhlich mechanism. The electronic cb and vb states interact with the lattice deformation due to the phonons resulting in energy shifts δE^c and δE^v of the band edges, according to:

$$\delta E^c_{\Lambda_1} = d^5_{1,o}(c)\frac{\hat{e}_l \cdot \boldsymbol{u}_{rel}}{2a_0}, \tag{6}$$

$$\delta E^v_{\Lambda_1} = d^5_{1,o}(v)\frac{\hat{e}_l \cdot \boldsymbol{u}_{rel}}{2a_0}. \tag{7}$$

Here the deformation potentials $d_1(c)$ and $d_1(v)$ of symmetry Λ_1 are the diagonal matrix elements of the deformation potential Hamiltonian [8] (two-band terms), $\hat{e}_l$ is a unit vector in the direction of one of the valleys $\Gamma \rightarrow \Lambda$ due to the energy minima at L in the vb and cb, $\boldsymbol{u}_{rel}$ is the lattice deformation, normalized to the lattice constant a_0. The vb splitting is changed and the vb states are mixed by the deformation potential d_3 (symmetry Λ_3) according to

$$\delta E_{\Lambda_3} = d^5_{3,0}\frac{|\hat{e}_l \times \hat{\boldsymbol{u}}_{rel}|}{a_0}. \tag{8}$$

This interband coupling clearly corresponds to the off-diagonal matrix elements of the deformation potential Hamiltonian [8], (three-band term). Both contributions are combined to form elements $a_{xy,yx}$ of the Raman tensor $\boldsymbol{R}$, (Raman polarizabilities) in backscattering geometry from a (001) surface. The two-band contributions show a very clear resonance at $E_1 + \hbar\Omega/2$ and $E_1 + \Delta_1 + \hbar\Omega/2$ (for $\hbar\Omega \leq \Gamma$), the three-band terms produce a broad resonance between E_1 and $E_1 + \Delta_1$. The electro-optic scattering by LO phonons due to the macroscopic field produced by these excitations is another contribution isomorphic to the deformation potential scattering [8,10,11]. This contribution is characterized by the Faust-Henry coefficient C [4,8], which is defined by the difference between the scattered amplitudes of LO phonons (a_{LO}) and TO phonons (a_{TO}), where it does not contribute:

$$\frac{a_{LO}}{a_{TO}} = 1 - \frac{\Omega^2_{LO} - \Omega^2_{TO}}{C\Omega^2_{TO}}. \tag{9}$$

$\Omega_{LO,TO}$ are the observed LO and TO phonon frequencies.

The main contribution of the macroscopic field to the Raman amplitude from a phonon is the Fröhlich intraband dipole-forbidden scattering. Its strength depends linearly on the phonon wavevector q and disappears for $q \rightarrow 0$. The intraband matrixelements of the Fröhlich Hamiltonian [8] evaluate the coupling of electronic states, which belong to the same pair (c,v) of bands. The Fröhlich mechanism gives rise to a strong resonance enhancement, as compared to the deformation potential, at $E_1 + \hbar\Omega/2$ even at intermediate damping constants. The contributions due to the different valleys near the L point have to be superimposed according to their orientation with respect to the incident electric field. The Raman tensor $\boldsymbol{R}_F$ for the Fröhlich term is in general nondiagonal. For backscattering on (001), (011), and (111) faces, however, $\boldsymbol{R}_F$ is diagonal with different numerical values for the tensor elements [9].

Another scattering mechanism with a diagonal scattering tensor $\boldsymbol{R}_{II}$ and competing with the Fröhlich intraband contribution is the impurity-induced scattering by LO phonons [9], where the intermediate electron-hole pair is scattered twice, first by the LO phonon via Fröhlich intraband interaction, and second by the coulombic potential of an ionized impurity. As the translational symmetry of the crystal is violated by an impurity, there is no wavevector ($\boldsymbol{q}$) conservation for this scattering mechanism and LO phonons of all $\boldsymbol{q}$ may contribute to the scattered amplitudes, the impurity carries away the surplus $\boldsymbol{q}$. Because of this fact, the impurity-induced scattering mechanism is, different from the previous intrinsic cases, an extrinsic effect and may contribute strongly under certain circumstances, although it appears in higher-order perturbation.

The total Raman intensity observed is a superposition of all contributions discussed above. Because the intrinsic mechanisms all lead to the same final state with a q value defined by the phonon under study, their amplitudes a_{ij} have to be added before squaring to obtain the intensity, quantum interference effects may occur. The extrinsic impurity-induced contribution is incoherent with the previous cases and has to be added by considering the intensities instead. By carefully adding all contributions and summing over all valleys the orientation-dependent Raman intensities given in Table 1 are derived for a (001) surface in backscattering geometry [9]. Similar relations can be obtained for the other orientations of crystal faces ((011), (111)), but are not given here. The isolated deformation potential contribution for TO phonons can be studied e.g. from a (011) surface under $\overline{z}(x,y)z$ or $\overline{z}(y,x)z$ polarizations. In the $z(y,x)z$ configuration from (001) only the dipole-allowed contribution will appear. The Faust-Henry coefficient C of HgTe can be determined by a comparison of the LO and TO data. Table 1 indicates the possibilities of constructive and destructive interference effects between different coherent contributions.

Table 1 Raman intensities from the (001) face for different parallel (e.g. (x,x), (x',x')) and crossed (x,y) polarization geometries; amplitudes a_i due to different scattering mechanisms: F: Fröhlich intraband, LO: Deformation potential and electro-optic, FI: Impurity induced.

(001) surface: $x=[1,0,0], y=[0,1,0], z=[0,0,1]$, $x'=\sqrt{\frac{1}{2}}[1,1,0], y'=\sqrt{\frac{1}{2}}[\overline{1},1,0]$	
$\overline{z}(x',x')z$	$\|a_F+a_{LO}\|^2+\|a_{FI}\|^2$
$\overline{z}(y',y')z$	$\|a_F-a_{LO}\|^2+\|a_{FI}\|^2$
$\overline{z}(x,x)z$ and $\overline{z}(y,y)z$	$\|a_F\|^2+\|a_{FI}\|^2$
$\overline{z}(y,x)z$ and $\overline{z}(x,y)z$	$\|a_{LO}\|^2$

It is evident from Table 1, that by various combinations of the intensities taken with different scattering geometries and from various oriented faces the contributions of the scattering processes discussed above can be separated. Details of this procedure will be published elsewhere [12] but results will be presented in Fig. 3 for the scattering by LO phonons from the (001) surface of HgTe.

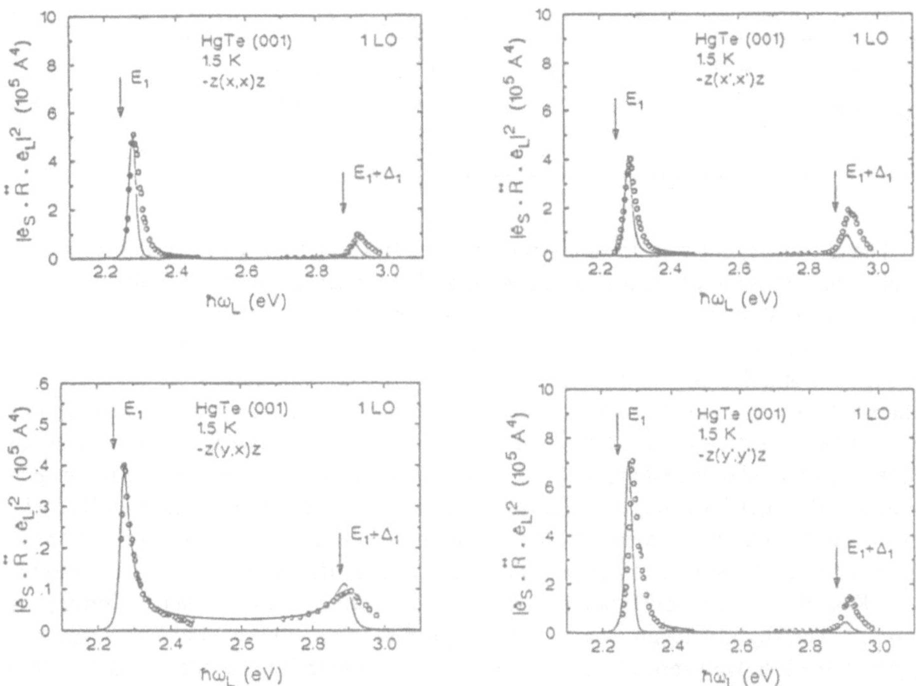

Figure 3 Experimental values of the absolute Raman scattering cross section of HgTe (001) in different scattering geometries (open circles) and theoretical calculations (fit, full curves).

The Raman resonances are shown as the squared Raman polarizabilities, the full lines give the results of the best fit to the experimental data using the theoretical expressions for the scattering cross sections and material parameters typical for HgTe [12]. The agreement is in general very satisfying, the larger experimental amplitudes at the $E_1 + \Delta_1$ frequency are probably due to uncertainties in the absorption correction. The free adjustable parameters in this fit were the relation between the dipole-forbidden and the interfering allowed amplitudes and the contribution of the impurity-induced intensity. A third parameter is a factor weighting a_F and adjusted to take the not explicitly considered electric field-induced scattering at the surface and the excitonic contributions neglected in the calculations into account. This factor is found to be 4.1 and is considerably smaller than in III-V compounds ($\approx$ 20 in GaSb [9]; $\approx$ 30 in ZnSe at the E_0 gap [13]), which indicates in fact the minor role of excitons played in HgTe in this context.

The results obtained for the deformation potential constants d_1 and d_3 are compiled in Table 2 together with other values for zincblende materials for comparison.

Table 2 Values of the deformation potential constants $d_{1,3}$ at the L point for HgTe ((111),[110])in comparison with other zincblende materials (GaP: [9]; CdTe, GaAs, GaP; calculated values only: [15]. p-doped HgTe(111), $T = 90$ K: $|d_1| = 24$ eV, $|d_3| = 21.6$ eV;[35].

	HgTe		CdTe	GaAs	GaSb	GaP
$d_{1,o}^5$ (eV)	−19.9	−21.9	−10.1	−10.6	−10.0	−6.3
$d_{3,o}^5$ (eV)	14.7	[13.0]	42.5	40.3	60.0	38.5

The d_1 values lie well in the range of other II-VI and III-V materials, while d_3, which describes the phonon-induced mixing of the valence bands, is smaller in HgTe as compared to other materials. This can be attributed to the inverted band structure of HgTe at the Γ point: The mixture of bands derived from an s- and a p-type partner as in HgTe is smaller than in materials with ordinary band structure, where only p-type states interact. It should be remembered, that values for other gaps in the same materials may differ considerably. The Faust-Henry coefficient C is positive in HgTe, contrary to most other zincblende materials. In CdTe, however, and in Te positive values have been found [14], which indicates that highly polarizable Te states contribute essentially to the electro-optic coupling in these materials. Impurity-induced incoherent contributions to the scattered intensity have not been detected. This indicates that at low temperatures (1.5 K in this experiment) ionized impurities do not exist, as the thermal energy is too low. The slightly asymmetric resonance profiles of the dipole-forbidden scattering profiles are less well reproduced by the calculations than for the allowed scattering. The scattering mechanisms discussed above are obviously not adequate to consider these effects. Surface contributions may be of importance, because the penetration depth of the exciting laser light is ≈ 200 Å in the regions of high absorption.

As Table 2 indicates, the knowledge of the interaction parameters d and others is only rudimentary in II-VI compounds. Additional information is available for the E_0 gap in wide-gap materials [15]. The role of impurity-induced scattering should be investigated by experiments at various temperatures. The resonance of 2LO phonons has also been discussed recently [15] but is not included here.

4 The Electronic Spin Flip in $(Cd_{1-x}Mn_x)Te$

$(Cd_{1-x}Mn_x)Te$ ($x \leq 0.7$) is a prominent example of a semimagnetic semiconductor. In these materials a strong exchange interaction exists between the extended electronic states at the band edge and the $3d^5$ states of the Mn^{2+} ions on regular lattice sites. These localized states lie well below the valence band and this interaction causes an unusually large spin splitting of valence and conduction bands. In mean field theory this splitting in $(Cd_{1-x}Mn_x)Te$ is proportional to the averaged Mn $3d$ spin orientation within the wavefunction range of the carrier. Therefore the Stokes shift of the spin flip Raman signal (SFRS) displays at low x values (below the spin-glass region) a Brillouin function-like behaviour with respect to magnetic field and temperature. It is a very sensitive probe of the Mn spin temperature and orientation. In the following discussion the detailed character of the carrier (free or bound electron) performing the spin flip and of the intermediate state of the Raman scattering process (free or bound exciton) will be determined.

The SFRS signal as well as the PRS treated in the next section are magnetic dipole transitions, where the angular momentum of the spin system is changed by one unit $\hbar$ ($\Delta m = \pm 1$). The Raman tensor is antisymmetric. To achieve this change, the polarization vectors of incident ($\boldsymbol{e}_L$) and scattered light ($\boldsymbol{e}_S$) must be orthogonally ((π, σ) or (σ, π)) polarized with each other, with one component perpendicular (σ) to the quantization axis, i.e. the direction of the magnetic field B ($\|$ [100] in our case), the other is parallel (π). This selection rule can be written as:

$$\boldsymbol{e}_L \times \boldsymbol{e}_S \times \boldsymbol{e}_B \neq 0 \tag{10}$$

for antisymmetric scattering to occur [16,17]. In a backscattering geometry, e.g. $\overline{z}(\sigma, \pi)z$, equ. (10) requires a Voigt scattering geometry ($B \perp z$). Equ. (10) is however valid only for the non-resonance case and may be violated in the case of resonance (see next section).

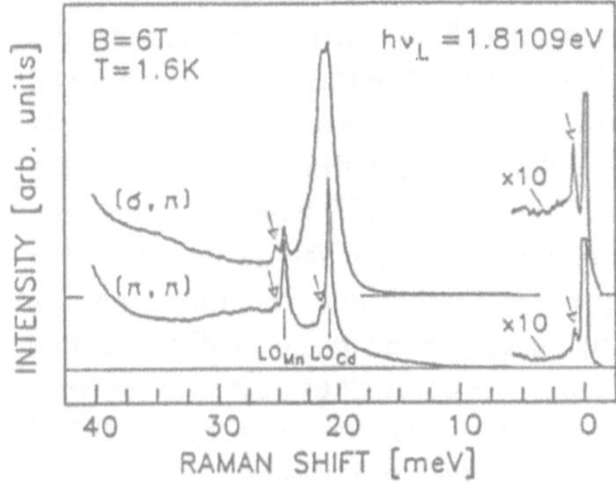

Figure 4
$\overline{z}(\sigma, \pi)z$ and $\overline{z}(\pi, \pi)z$ Raman spectra of a n-doped (In) $Cd_{0.865}Mn_{0.135}Te$ epilayer, 1.6 μm thick, (001)-oriented on a $Cd_{0.96}Zn_{0.04}Te$ substrate at $B = 6$ T, $T = 1.8$ K with excitation in the energy range of the spin split excitonic transitions. The baseline of the (σ, π) spectrum, which exhibits the SFRS signal near 20 meV, is shifted for clarity. The intensity increase at shifts ≥ 40 meV is due to intense bandedge luminescence. The broad background peak in the (π, π) spectrum near 27 meV is caused by the recombination of light hole (lh) excitons. The spectra also show the 1LO phonon transitions and the paramagnetic resonance signal (PRS).

In Fig. 4 a typical Raman spectrum of $(Cd_{1-x}Mn_x)Te$ is depicted [18]. As expected, the SFRS signal is observed only in the (σ, π) spectrum. At $B = 6$ T the SFRS signal is almost degenerate with the LO_{Cd} phonon in this two-mode material. As a phonon companion and close to the Rayleigh line (marked by arrows) the PRS signal is found (see next section).

In Fig. 5 the Brillouin-type B dependence of the SFRS signal is shown, which agrees well with the spin splitting of the conduction band as is measured e.g. by PLE.

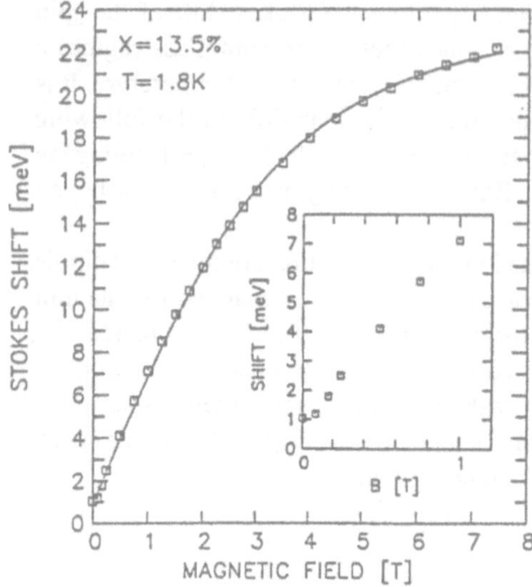

Figure 5
Stokes shift of the SFRS line shown in Fig. 4 with magnetic field B. The effect of the magnetic polaron is demonstrated in the insert. The solid line is a fit using the formula of Peterson et al. [16], which goes beyond mean field theory by considering the effects of magnetic fluctuations.

The insert demonstrates the deviation from the Brillouin behaviour at low magnetic fields due to the formation of an electron magnetic polaron. The magnetic polaron is a complex where the spin of a carrier and the localized magnetic moments of the Mn^{2+} $3d$ states interacting with it tend to be aligned with each other due to exchange interaction. This results in a decrease of the total system energy and in a finite splitting of the electron states of ≈ 1 meV even at $B = 0$ T. Thermodynamically at $T = 2$ K an electron magnetic polaron should only be stable if the carrier is trapped by Coulomb interaction with a donor or acceptor, respectively, in addition to the rather weak magnetic localization [19,20]. Besides the observed spin flip of an electron, the appearance of a residual spin splitting at $B = 0$ T of about 1 meV displaying magnetic polaron formation indicates, that the electron changing its spin should be trapped at a *donor* by Coulomb interaction. This is in agreement with the observation, that the intensity of the observed signal is proportional to the laser intensity, i.e. no photoexcited carriers contribute to the spin flip scattering. This is in contrast to observations with *p*- doped $(Cd_{1-x}Mn_x)Te$ samples [21], where a superlinear behaviour of the SFRS intensity as a function of the excitation power indicates an attribution of this spin flip line to photoexcited electrons trapped at ionized donors (here $N_A - N_D = 5 \cdot 10^{16}\ \mathrm{cm}^{-3}$).

The halfwidth of the spin flip Raman line displays a constant width at fields $B \leq 1$ T due to the temporal fluctuations of the magnetization in the magnetic polaron regime. At larger fields, where the magnetic polaron effect is suppressed because of the spin alignment in the external field, the width increases linearly with B, mostly because of alloy fluctuations of the total number of Mn spins interacting with the electrons bound at the statistically distributed donors.

A schematic representation of the spin splittings in the Γ_8 valence and Γ_6 conduction

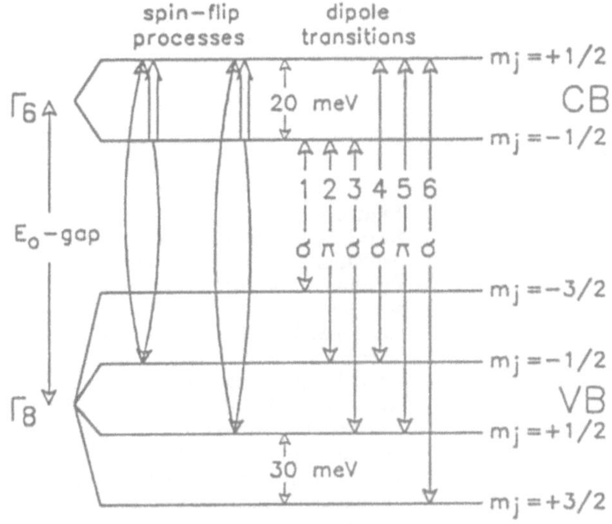

Figure 6
Schematic spin splitting of the Γ_8 valence and Γ_6 conduction band states. Dipole transitions, allowed in Voigt configuration, are shown with their respective polarizations $((\sigma), (\pi))$. On the left side the two Raman processes contributing to the SFRS on Γ_6 are indicated.

band states is shown in Fig. 6. Dipole transitions, which are allowed in Voigt geometry, are labelled by numbers 1 to 6 and characterized by their polarizations σ and π. Dipole selection rules allow the SFRS signal in resonance in (π, σ) and in (σ, π) geometry. Therefore scattering at the spin of an Γ_6 conduction electron (j = 1/2) is resonantly excited only via the $m_j = 1/2$ (light hole) valence band components as intermediate states. The two resulting processes for Stokes scattering on Γ_6 electrons are shown on the left side of Fig. (6): The exciting photon creates an electron hole pair with the electron in the $m_j = 1/2$ excited spin state and the hole in one of the $m_j = \pm 1/2$ states, where the sign of the hole state depends on the scattering geometry. In the second step this hole recombines with an electron in the occupied $m_j = -1/2$ conduction band state by emission of scattered light. The result is a spin flip of an electron from $m_j = -1/2$ to $m_j = +1/2$ while the valence band is left unchanged. The scattered light is Stokes shifted by the spin splitting of the electron states for any exciting laser energy. Due to the resonance denominators occuring in (3) one can expect maximum scattering probability, if the energy of the laser photon is equal to the $4(\sigma)$ or the $5(\pi)$ transition (Fig. (6)) depending on the geometry. At resonance the energy of the scattered light then has to be equal to the $2(\pi)$ or the $3(\sigma)$ transition respectively.

Fig. 7 shows in its upper part the scattering efficiency of the SFRS excitation plotted as a function of the energy of scattered light. A clear resonance is observed in both geometries. To characterize the dipole transitions involved, the lower parts in Fig. 7 display luminescence and PLE spectra measured under the same experimental conditions (B, T). The distinct features in the PLE spectra at 1.76, 1.79 and 1.81 eV are attributed to the creation of free excitons and correspond the $1(\sigma)$, $2(\pi)$ and $3(\sigma)$ transitions in Fig. 6. Resonance for the SFRS is observed in agreement with the dipole selection rules, if the energy of the scattered light is equal to the transitions $2(\pi)$ and

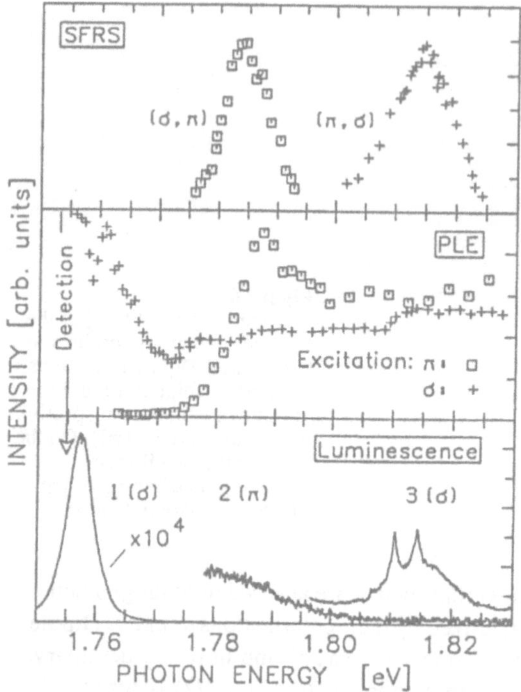

Figure 7
SFRS, PLE (photo-luminescence excitation), and luminescence as a function of the quantum energy of the scattered (exciting) light for a $Cd_{0.865}Mn_{0.135}Te$ epilayer at $B = 6$ T, $T = 1.8$ K, plotted on the same energy scale. Luminescence shows the recombination of the $1(\sigma)$ hh exciton and the much weaker recombination of $2(\pi)$ and $3(\sigma)$ lh excitons. The PLE spectra were recorded on the $1\,(\sigma)$ luminescence (arrow), while the exciting laser energy was increased. The features displayed correspond to the lowest σ and π exciton transitions. SFRS shows resonance if the energy of the scattered light is equal to the $2(\pi)$ or the $3(\sigma)$ excitonic transition, depending on the scattering geometry.

$3(\sigma)$. As the exciton binding energy in CdTe is 10 meV, transitions between states of free electron-hole pairs and exciton transitions can be clearly separated. From the good agreement between the position of the excitonic maxima in PLE and of the resonance for SFRS efficiency the conclusion can be drawn, that excitons serve as intermediate states for the SFRS process [16].

In the case of SFRS on donor bound electrons the photoexcited exciton with electron spin $m_j = +1/2$ is instantaneously interacting with the $m_j = -1/2$ donor electron. This intermediate state of the SFRS process therefore should be a donor bound exciton (D^0, X), i.e. a localized excited state of the neutral donor complex. The (D^0, X) involved in SFRS can be considered as a diatomic molecule-like system with the donor and the exciton's hole component as nuclei, while the binding energy stems from two s-like electrons in a singlet state. From this model one can expect the resonance of the scattering process to be Stokes shifted relatively to the free exciton transition by the binding energy of this pseudo-molecule, i.e. the localization energy of the exciton at the donor. For CdTe the localization energy of excitons at substitutional donors is known from luminescence to be 3.3 to 3.6 meV, depending slightly on the type of the donor [22]. In Voigt geometry a precise determination of the $2(\pi)$ lh free exciton energy is possible , e.g. by PLE (Fig. 8). On the other hand the SFRS resonances are rather distinct and permit a very accurate comparison between the two excitation energies. Fig. 8 shows that the Raman resonance is Stokes shifted relative to the free exciton transition by 3 to 4 meV in good agreement with the localization energy of an exciton at a substitutional neutral donor. This binding energy is not affected by the magnetic field as it is mainly due to the exchange energy of the electronic singlet.

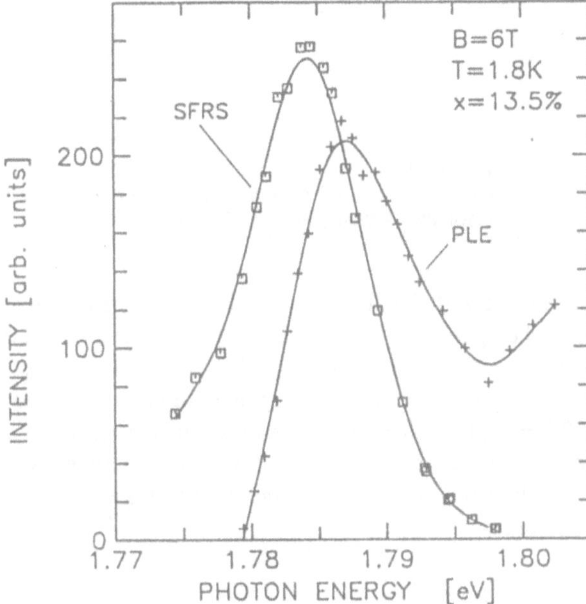

Figure 8
Intensity of the SFRS in $Cd_{1-x}Mn_xTe$ in (σ, π) geometry as a function of scattered photon energy (squares) and PLE structure (crosses) of the free exciton absorption at the $2(\pi)$ transition.

It should be realized that the singlet state of the (D^0, X) complex may not be a stable state as triplet states with lower energy may exist, i.e. the lifetime of this state may be in the picosecond range due to the strong exchange interaction, which allows fast spin transfer to the Mn^{2+} $3d$ states. Due to the very fast time scale of the scattering process, however, it does not matter if the intermediate state is thermodynamically stable.

In conclusion, it appears that a donor bound exciton serves as intermediate state in the SFRS Raman process. This has to be compared with the extensively studied resonance behaviour of phonons, which also proves the importance of excitonic contributions to the observed resonance profiles [4, 22, 23, 24].

5 Optically Detected Paramagnetic Resonance of Mn^{2+} in $Cd_{1-x}Mn_xTe$

The narrow Raman line close to the Rayleigh (laser) line in Fig. 4, which also occurs as a LO phonon companion, is the paramagnetic resonance signal (PRS) of the localized $3d$ electrons in the $3d^5$ configuration of Mn^{2+} ($L = 0$, $S = 5/2$, ^6S). This

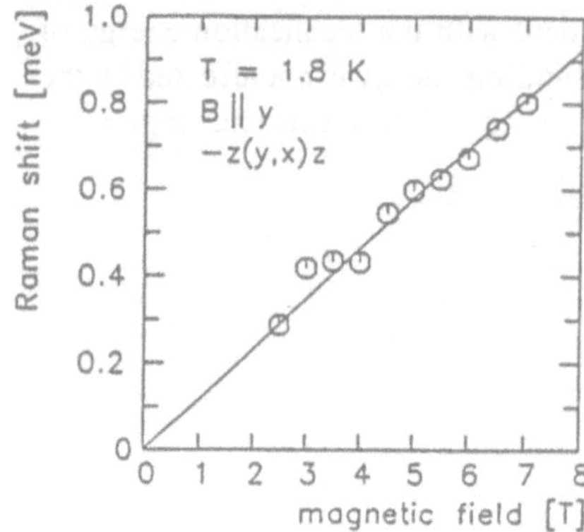

Figure 9
Optically detected Zeeman shift of the paramagnetic resonance signal (PRS) in (Cd,Mn)Te.

becomes evident from the shift of this line with B (Fig. 9), which is linear, extrapolates to zero Raman shift with $B \to 0$, and displays a g factor of 2.00 ± 0.1, as determined optically, which compares favourably with the more precise EPR value: $g = 2.010$ [26]. The experiments have been performed using a MBE-grown, nominally undoped epilayer ($d = 1.6\ \mu$m , $x = 0.135 \pm 0.01$). Other centers with a g factor close to 2 do not exist in this material. Free electrons or holes are subject to the strong exchange interaction (see previous section) with huge magnetic reorientation energies, Mn^{1+} complexes have clearly different g values.

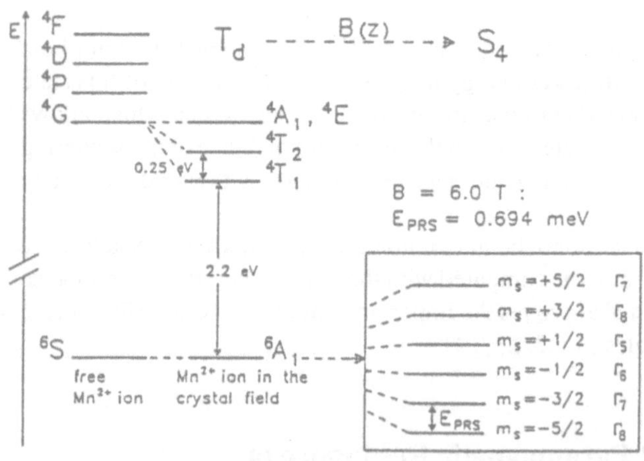

Figure 10
Energy levels of $3d^5$ (Mn^{2+}) on a Cd site with symmetry T_d of $Cd_{1-x}Mn_xTe$ and splitting of the ground state ^6S in an external magnetic field $B||[100]$. The double-group irreducible representations $\Gamma_5 - \Gamma_8$ of S_4 ($\bar{4}$) are given according to the tables of Koster et al. [27].

In Fig. 10 the excited levels of the $3d^5$ configuration are schematically shown, with the splitting of the ^6S groundstate in an external magnetic field $B||[100]$, which reduces the site symmetry of Mn^{2+} on a Cd site from T_d to S_4 ($\bar{4}$).The polarization selection rules for magnetic dipole Raman scattering can be obtained by the usual group-theoretical procedure: A magnetic dipole moment R ($\Delta m = \pm 1$) transforms according to Γ_4 of

T_d, in S_4 according to Γ_1 (R_z, antisymmetric) and $\Gamma_3, \Gamma_4(R_x, R_y)$, $B||z$ [27]. Quadrupolar transitions ($\Delta m = 0, \pm 2$) transform according to Γ_3, Γ_5 (T_d) or to Γ_1 (symmetric), $\Gamma_2, \Gamma_3, \Gamma_4$ of S_4. The representations of the (groundstate) ^6S multiplet are indicated in Fig. 10, the transitions between neighbouring levels transform in S_4 according to Γ_3 ($\Delta m = +1, (-3)$), Γ_4 ($\Delta m = -1, (+3)$), according to Γ_2 for $\Delta m_J = \pm 2$), and to Γ_1 for $\Delta m_J = 0, (\pm 4)$. The $\Delta m = \pm 1$ transitions are thus observed in Raman scattering for the $\overline{z}(yx, xy)z$ orientations, where now the direction of y is parallel to B and z is the growth direction of the epilayer, i.e. in a backscattering experiment only Voigt geometry should be applicable for the observation of the PRS. The quadrupolar transitions ($\Delta m = 0, \pm 2$) can be observed both in Voigt geometry $\{\overline{z}(x, x)z, (\Gamma_1, \Delta m = 0)$ or $(\Gamma_2, \Delta m = \pm 2); \overline{z}(y, y)z, (\Gamma_1)\}$ and in Faraday configuration ($B||z$) using circularly polarized radiation for the incoming and the outgoing radiation $\overline{z}(\sigma^+, \sigma^+)z$ and $\overline{z}(\sigma^-, \sigma^-)z, (\Delta m = 0)$ or $\overline{z}(\sigma^+, \sigma^-)z, z(\sigma^-, \sigma^+)z, (\Delta m = \pm 2)$. The selection rule (10) should be violated in resonance and magnetic spin flips should occur in other polarization orientations than predicted by (10).

In Fig. 4 the PRS is indicated by arrows. Evidently the PRS is observed in both (σ, π) and (π, π) spectra, i.e. in the "forbidden" (10)) orientation, contrary to the other magnetic dipole transition, the electronic spin flip, which obeys rule (10). The width of the PRS is determined by instrument resolution at all B values. These results agree well with those of Petrou et al. [28].

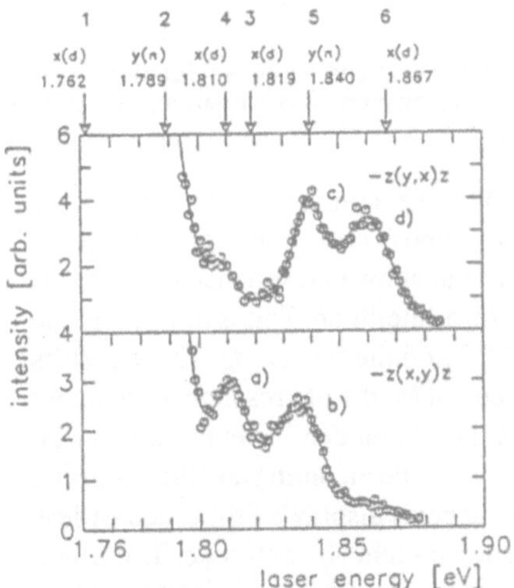

Figure 11 Intensities of the PRS as a function of the exciting laser energies in "allowed" (crossed) polarizations at $B = 6$ T, $B||[010]$, (Voigt), in the plane of the epilayer. The numbers, energies (eV), and arrows on the top refer to the transitions shown in Fig. 6. $\Delta m_S = 1$ transitions.

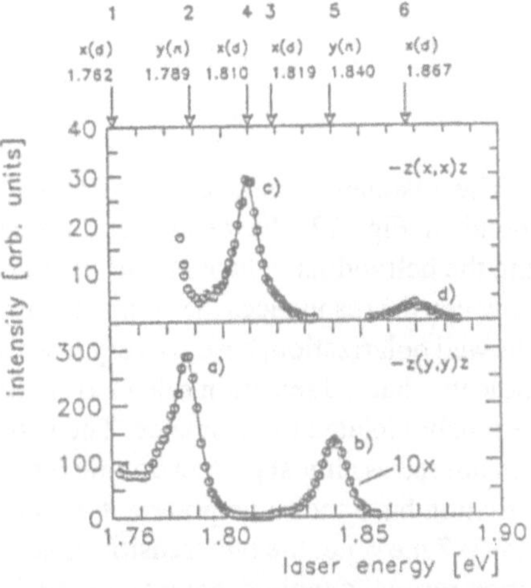

Figure 12 Same as in Fig. 11, but in forbidden (parallel) polarization (Voigt). $\Delta m_S = 0$ transitions.

Figs. 11 and 12 display the intensities of the PRS near the Rayleigh line as a function of the quantum energy of the exciting laser in Voigt geometry for two orthogonal polarizations ("allowed" (see (10), Fig. 11) and parallel polarization ("forbidden" by (10)), (Fig. 12). Distinct extrema of the scattering cross section are found in both orientations, which can be easily related to the various excitonic transitions indicated in Fig. 6. This fact indicates the role of excitonic states as intermediate states also in this scattering process. In Fig. 13 the resonances for Faraday geometry are depicted. Resonances are observed for identical circular polarizations (σ^+, σ^+) and (σ^-, σ^-), $(\Delta m = 0)$, but not for orthogonal polarizations (σ^+, σ^-), $(\Delta m = \pm 2)$ etc.

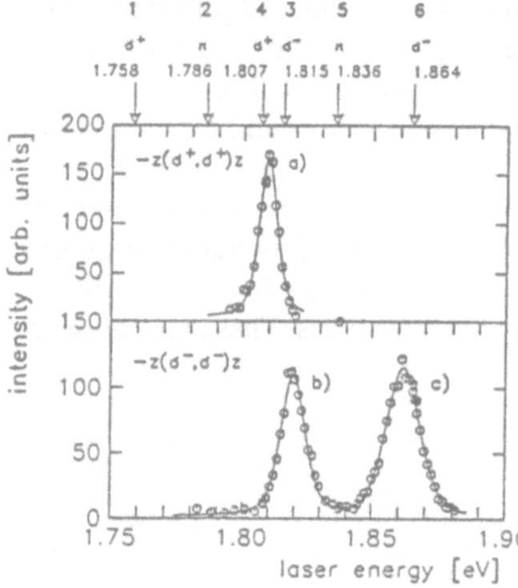

Figure 13
Same as in Figs. 11 and 12, but in Faraday orientation, using circular polarization. $\Delta m_S = 0$ transitions.

The resonance enhancements and the halfwidths are comparable to the results found in Fig. 12, the resonances are stronger by almost two orders of magnitude and the halfwidths reduced by $\approx 2^{1/2}$ as referred to the allowed resonances in Fig. 11. Between the resonances the signal drops to almost zero amplitude. This is not true in the allowed polarization, here at every laser energy a PRS of finite size exists. These results indicate, that the selection rule (10) is obviously obeyed for the off-resonance case, but seriously violated in resonance. The resonances in the "forbidden" orientations can be considered as almost perfect double resonances, because the incoming and the outgoing radiation have the same polarization and their frequencies display a difference of less than 0.7 meV, i.e. the two transitions start and end at the same vb or cb state. The much larger resonance enhancements for the forbidden resonances support this interpretation (see sect. 2). It should be realized, that the allowed resonances correspond to spin flip processes with $\Delta m = \pm 1$ (one unit of angular momentum transfer between crystal and light field), while the forbidden transitions are $\Delta m = 0$ processes (no angular momentum transfer), but a $\Delta m_S = +1$ transition occurs in the Mn^{2+} ground state, as is obvious from the observed B-dependent shift.

For a deeper understanding of the optically detected PRS the interaction between

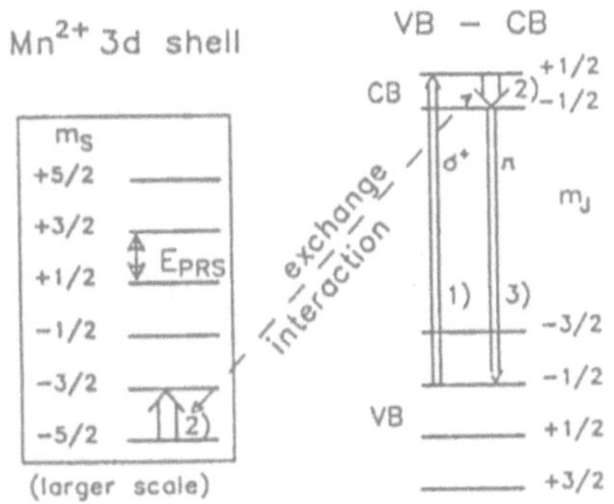

Figure 14
Mechanism of Stokes scattering of PRS via exchange coupling between the spin of a cb electron and the Mn^{2+}-spin in its ground state. σ^+ and π indicate virtual electric dipole transitions due to the incoming laser radiation (1) and the outgoing scattered radiation (3). Processes (2) indicate the simultaneous spin flips of the electronic (cb) and the localized spins. A corresponding process can be constructed for a spin flip in the vb (hole process, see Fig. 15). Energies are not to scale.[28]

the magnetic moment of the Mn^{2+} ion and the scattered light wave must be studied. As the observed resonances of the scattering cross section are coinciding with the excitonic interband transitions, the bandedge states of the vb and cb are involved in the scattering process. This proves that for the spin flip of the Mn^{2+} an exchange process as indicated in Fig. 14 has to be considered as proposed by Petrou et al. [28]. In the example depicted in Fig. 14 a virtual transition (real in the case of resonance) from the $m = -1/2)$ state of the vb to the $(m = +1/2)$ state of the cb is induced by absorption of a σ^+ photon, followed by an electronic spin flip to the $(m = -1/2)$ state of the cb and subsequently by a virtual recombination of the electron fom the $(-1/2)$ cb state to the $(-1/2)$ vb state with π polarization. The virtual spin flip $(\Delta m = -1)$ is coupled by the sp-$3d$ exchange interaction with a real spin flip $(\Delta m_S = +1)$ of the $3d^5$ shell. An equivalent process can be constructed for a virtual spin flip of a hole in the vb, exchange-coupled to the $3d^5$ shell.

The interaction Hamiltonian:

$$H_{el} = \sum_n [J_e(\boldsymbol{r}_e - \boldsymbol{R}_n)\boldsymbol{S}_n \cdot \boldsymbol{s}_e + J_h(\boldsymbol{r}_h - \boldsymbol{R}_n)\boldsymbol{S}_n \cdot \boldsymbol{s}_h] \quad (11)$$

with $\boldsymbol{r}_e, \boldsymbol{r}_h$; $\boldsymbol{s}_e, \boldsymbol{s}_h$ the position and spin operators of electrons and holes, respectively. $\boldsymbol{R}_n$, $\boldsymbol{S}_n$ refer to the n-th Mn^{2+} ion, $J_{e,h}$ are the exchange integrals. This Hamiltonian can be decomposed into shift operators, comprising $\boldsymbol{S}_n^+ \cdot \boldsymbol{s}_e^-$, $\boldsymbol{S}_n^- \cdot \boldsymbol{s}_e^+$. Other processes than shown in Fig. 14 can be conceived, starting from other m_J levels of the vb or including spin flips of holes in the vb. The comparison with the experimental results indicates, that the observed resonances in the allowed (crossed) polarizations coincide well with the incoming or outgoing resonances expected from Fig. 6. All observed spin flips can be interpreted as hole processes (the virtual spin flip occurs in the vb), but only a few also as electron processes, alternatively. This predominance of the hole processes is in agreement with the ratio of the exchange integrals $J_e/J_h = -0.25$ [29], i.e. hole processes display the higher efficiencies.

The appearance of the forbidden (doubly-resonant) PRS spectra and the shapes of the resonances suggest, that again an exchange mechanism similar to the one just discussed is most probable. Interaction mechanisms, which are completely different

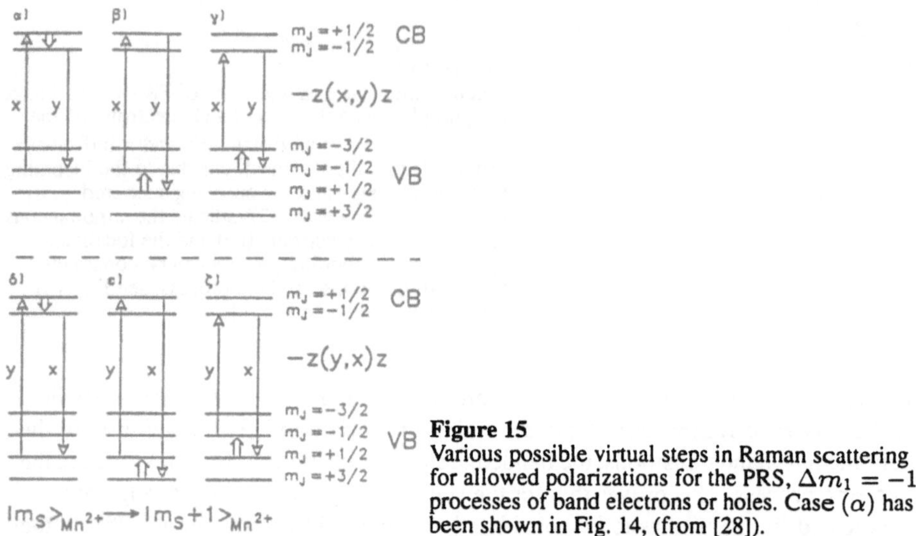

Figure 15
Various possible virtual steps in Raman scattering for allowed polarizations for the PRS, $\Delta m_1 = -1$ processes of band electrons or holes. Case (α) has been shown in Fig. 14, (from [28]).

from the previously discussed, have to be considered with caution. Examples of the virtual transitions in and between the vb and cb occuring here are shown in Fig. 15 for the specific case of (σ^+, σ^+) polarization. With these mechanisms one up spin flip and one down spin flip are connected with a total $\Delta m = 0$, but involving two exchange processes. Such a process will occur, if besides the Mn^{2+} spin there exists another spin excitation with an orientational energy undetectably small in Raman spectroscopy, but which balances the angular momentum to zero and couples with the charge carriers either in the vb or cb. Each of the forbidden spectra either in Voigt or in Faraday orientation can be interpreted by doubly resonant processes. Only in a few cases singly resonant transitions ($\Delta m = \pm 2$) may occur in superposition.

For an interpretation of the forbidden transitions and the apparent violation of the conservation of angular momentum in these experiments there exist several possibilities, which will be shortly discussed in the following: First, the symmetry may be lowered by the application of the magnetic field $B||[100]$ or locally by perturbations due to lattice defects. $m(\bmod 4)$ in S_4 should not be considered as a good quantum number. The symmetry discussion given above negates the interpretation with respect to B, local distortions of the lattice should lead to a breakdown of selection rules for all transitions observed, independent of B, the electronic spin flip included.

Second, there is one type of lattice defects in semimagnetics, which deserves a closer inspection. At a concentration $x \approx 0.1$ of the Mn^{2+} ions as in these experiments, there exist different types of exchange coupled clusters of Mn^{2+} ions in (Cd,Mn)Te (pairs, open and closed triangles etc.). Their magnetic excitation energies, which differ by one or several units of angular momentum from each other, are known from experiments and have been studied theoretically [30,31]. These results show, that clusters cannot be the missing partner in the present problem to compensate the total angular momentum to $\Delta m = 0$ in the transition, since their excitation energies should be easily observable in the Raman spectra, either as additional lines or at least as a broadening of the PRS in the case of larger clusters. Both features are not observed. Clusters can also act as local defects, lowering the symmetry of neighbouring (regular) lattice sites. The argument given above applies however also in this case.

Finally, it appears most probably, that the nuclei of ^{55}Mn with an abundance of 100 % and a nuclear spin $I = 5/2$ and the spins of the Cd isotopes with $I > 0$ ($\approx$ 25 % abundance) are involved via hyperfine interaction. Te has two isotopes with a nuclear spin $I = 1/2$ occuring with a total abundance of 8 %. Charge carriers and the nuclei may interact in two ways: First indirectly via the $3d$ shell, i.e. the band electron or the hole flips the $3d$ spin according to (11), then a hyperfine interaction according to

$$H_{HF} = A\boldsymbol{S} \cdot \boldsymbol{I} \tag{12}$$

provides the nuclear reorientation in the same Mn^{2+} ion with $\Delta m_S = \pm 1, \Delta m_I = \mp 1$. This is a process appearing in higher order approximation and should be less probable than a direct interaction of the charge carriers and all nuclei with $I \geq 1/2$, where the s electrons of the cb interact via the Fermi contact term A' (12) with the nuclei. In this case A' is essentially determined by the gyromagnetic ratio of the nucleus and by $|\Phi(R)|^2$, where $\Phi(R)$ is the amplitude of the lattice-periodic part of the electronic Bloch function at the nuclear site. On the other hand, the electrons in the cb and the holes in the vb may couple with the nuclei by magnetic dipole interaction, which is in most cases weaker than the exchange mechanism. This dipolar interaction occurs, however, in a lower order of perturbation, which may compensate for its inherent weakness. In Figs. 16, 17 the exchange interaction is schematically shown for the special case of the ^{55}Mn nucleus.

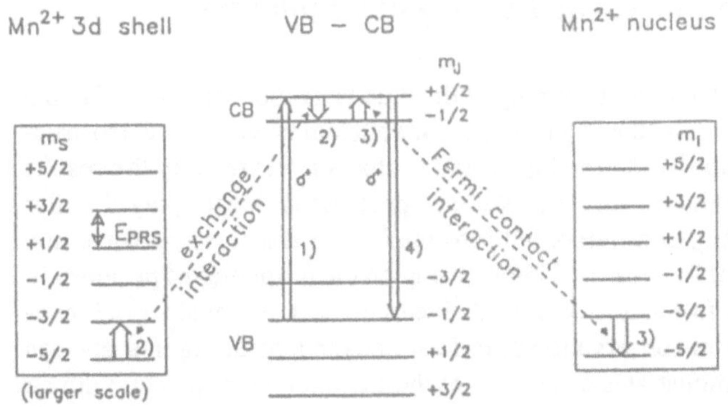

Figure 16
Example of simultaneous interactions between virtual transitions in the cb with the $3d$ shell of Mn^{2+} via exchange interaction and with the ^{55}Mn nucleus via contact interaction. Energies not in scale, $B > 0$.

There are numerous evidences for the effectiveness of the exchange interaction in semiconductors, especially in the field of optical orientation of electrons and nuclei, where the observed nuclear relaxation times are governed by the coupling to the

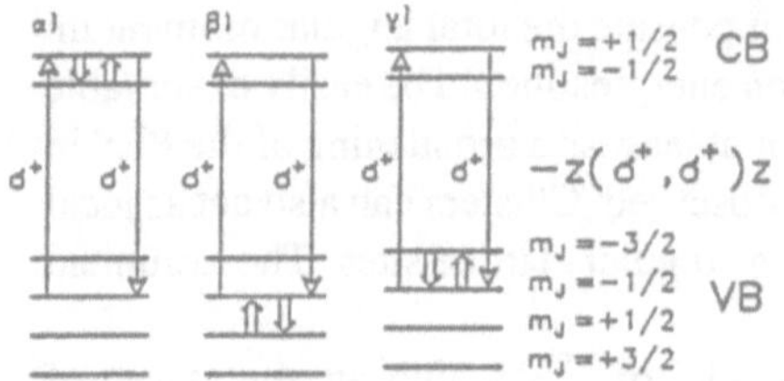

Figure 17
Examples of possible virtual transitions in the forbidden ($\Delta m = 0$) doubly resonant transitions (σ^+, σ^+, Faraday configuration, see Fig. 13).

conduction electrons. In fact the experimental procedure in Raman spectroscopy of shining linear or circular polarized laser light on the sample in resonance with electronic transitions near the band edge is known to produce an easily detectable dynamic nuclear polarization in semiconductors without paramagnetic centers (e.g. in GaAs).

The observations presented in this section fully support the model of Petrou et al. [28] for the detection of the PRS by Raman spectroscopy. They are the first Raman resonance experiments in this field, which fully cover all allowed PRS resonances. Resonances in forbidden polarizations have not been detected before but also exist, they are in fact double resonances and have to be interpreted as transitions with a zero change of total angular momentum ($\Delta m = 0$), however the observed (Raman) energy shift indicates a transition $\Delta m_S = 1$. The partner, which contributes the compensating angular momentum but does not increase the magnetic orientation energy within the precision of a Raman experiment, is most probably a nucleus (^{55}Mn, ^{111}Cd, ^{113}Cd, ^{123}Te, ^{125}Te) with $I > 0$. If this hypothesis can be validated, a new series of experiments concerning hyperfine interactions in this semimagnetic material can be started. On the other hand, the contribution of magnetic dipole interaction has to be studied both experimentally and by theoretical calculations.

6 Raman Scattering in Heavily Doped ZnSe Epilayers

The recent interest in ZnSe technology aiming at the blue laser diode has stimulated a large number of investigations on the chances to grow epitaxial layers and heterostructures of improved quality, both n- and p-doped, and to characterize reliably the results of such efforts [32, 33]. Luminescence and Raman spectroscopy again play a major role in this respect, providing both important information for the crystal grower and giving access to processes of fundamental interest, e.g. on the properties of phonons in weakly disordered crystals. In a disordered solid, due to the lack of translational invariance, the crystal momentum conservation criteria are relaxed, resulting in a phonon Raman spectrum which demonstrates a decrease of the LO phonon frequency with an increasing concentration c of the dopant accompanied by an asymmetric broadening of the line [34].

In an ideal crystal, the region of correlated motions of the atoms in a certain optical phonon mode is large. When the crystal is doped, the range of correlated motions decreases due to scattering induced by the defects. In this case the phonon mode is not located at the Γ point of the BZ alone but probes a region around this center point, where the frequencies for finite q are usually lower due to dispersion. Consequently, a shift of the line center to lower frequencies and an asymmetric shape with a smaller slope on the low energy tail of the band will appear. The diameter of the correlated regions will be roughly proportional to the average distance between crystal defects and the phonon anomalies will grow with c.

In Fig. 18 the Raman spectra in the region of the LO phonon of an epitaxial layer ($d \approx 1\ \mu$m) of n-doped ZnSe:Cl grown on GaAs (100) substrates is shown for three Cl concentrations close to 10^{19} cm^{-3} [33]. SIMS measurements have shown that the concentration profile of chlorine is not constant but make the existence of a region of larger c values at or near the surface very probable. The average densities of charge-carriers have been determined by van der Pauw measurements. The spectra show a narrow line close to the (bulk) frequency of the LO phonon, accompanied on the low frequency side ($\Delta\nu \approx 20$ cm^{-1}) by a broad band, which increases in intensity with c. A closer inspection of the LO phonon lines clearly displays the expected asymmetry, which is growing with c. Similar results have also been obtained in heavily doped CdTe. The Raman resonance behaviour of both excitations has been studied: The LO phonon displays a resonance of the scattering cross section close to the band edge of ZnSe, as expected. The broad companion, however, displays a resonance shifted by about 400 cm^{-1} to higher energies. The eigenfrequencies of both excitations are not constant when the exciting laser sweeps through the resonance, but decrease by ≈ 20 cm^{-1} with increasing quantum energy.

The observed behaviour can be interpreted assuming a c profile of chlorine with c increasing toward the surface and a region of high c near the surface, as the SIMS results suggest. The broad feature is due to the LO phonon branch in this high-c region close to the surface with a strongly reduced phonon correlation sphere in real space, involving the whole LO branch in q space, however with decreasing weight of frequencies at large q values according to a Gaussian distribution [34]. The narrow LO band is excited in low-c regions underneath the surface. The decrease of eigenfrequencies is due to the increasing absorption of the laser radiation by bandedge states when entering the resonance region from lower quantum energies. With increasing absorption the regions, where the main contributions to the scattered intensities arise, are shifting toward the surface and the c-dependent effects will grow. A quantitative analysis along the lines indicated above give a surprisingly good agreement of the phonon correlation length fitted with the values expected from the Cl concentrations. Recent etching experiments have corrobated the existence of a c profile with a region of high chlorine concentration near the surface.

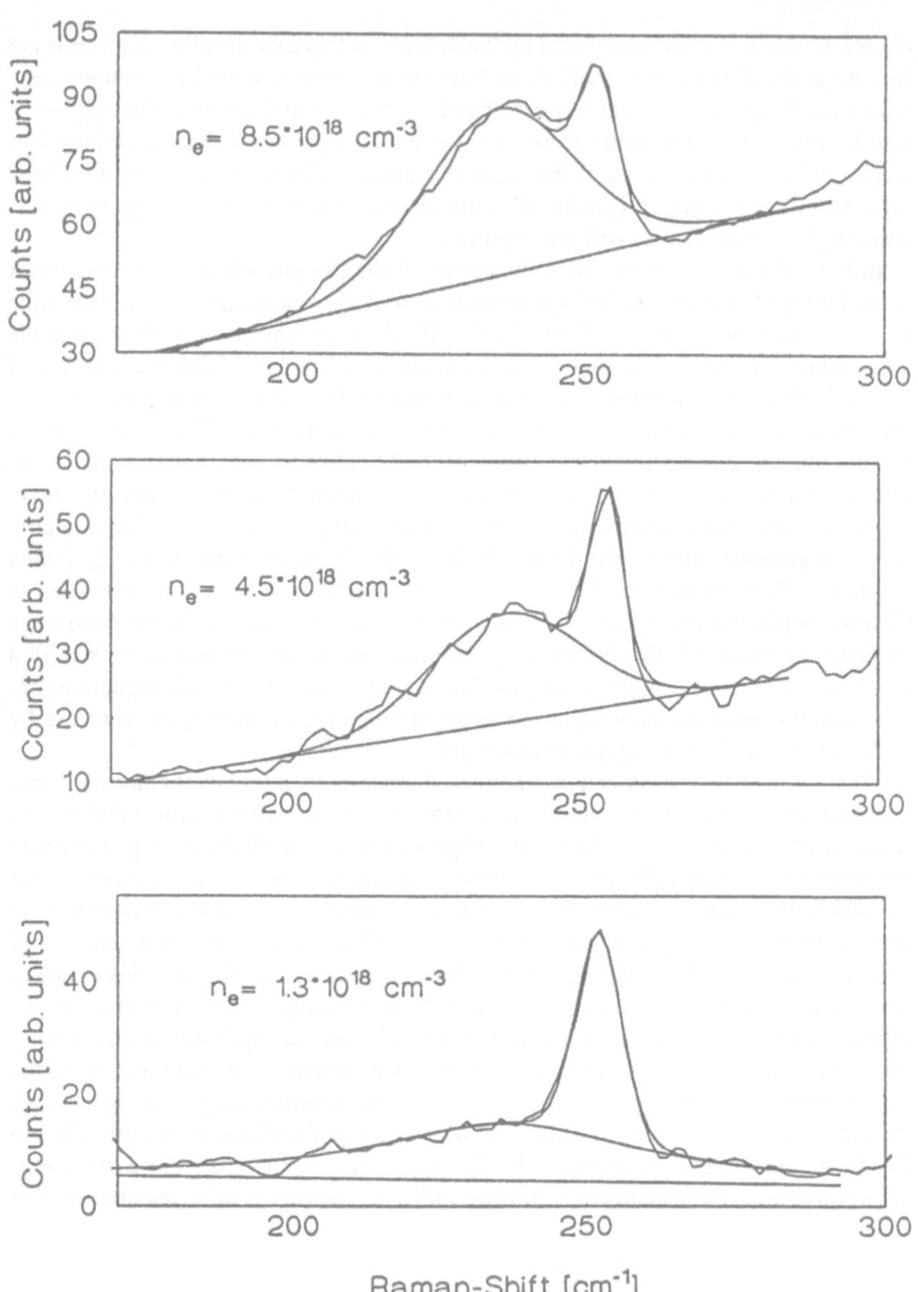

Figure 18 Raman spectra at $T = 1.8$ K of n-doped ZnSe:Cl in the region of the LO phonon. The electron concentrations given have been determined by the van der Pauw method.

7 Conclusions

The physics and the technology of II-VI compounds are rapidly growing fields with promising developments in the future. Classical spectroscopic techniques as Raman spectroscopy and related methods will contribute essentially to the expected progress. The fields of future developments can be anticipated by comparing the present state of II-VI research with the matured area of III-V materials. Future technological achievements in producing purer materials, more sophisticated techniques of crystal growing and detailed experience in overcoming spontaneously arising lattice defects will bring II-VI structures closer to the perfection, which is now standard in the III-V field. Heterostructures of all kinds with better perfection of interfaces and more flexibility in the methods for doping at high concentrations will certainly become available.

The field of diluted magnetic semiconductors, which is an exciting speciality of the II-VI group of materials, will benefit widely from these developments. Here the fundamental research will find interesting and challenging new tasks. Magnetism in thin layers intercalated between magnetically inert materials, new systems of quantumwell structures with the magnetic ions located in the well, as is available in (Cd,Mg)Te-(Cd,Mn)Te structures, semimagnetics with a changing valency of the magnetic ions ($Fe^{2+} \leftrightarrow Fe^{3+}$) will impose new problems to be attacked in the foreseeable future.

Acknowledgements

The author is indebted to R. Atzmüller, M. Dahl, M. Hirsch, M. Kling, J. Kraus, R. Meyer, M. Rösch, J. Stühler, for providing most of the material compiled. Ch. Becker and A. Waag, all from Würzburg, have grown the samples used. The Deutsche Forschungsgemeinschaft has supported these investigations in various projects.

Bibliography

[1] K. Birus, in Ergebnisse der exakten Naturwissensch., Vol. 20, ed. by F. Trendelenburg, Springer, Berlin (1942). G. Heiland and E. Mollwo, in Solid State Physics, Vol. 8, ed. by F. Seitz and D. Turnbull, Academic Press, New York and London, 1959. R. Dornhaus and G. Nimtz, Springer Tracts in Mod. Physics, Solid State Phys., Vol. 78, ed. by G. Höhler, p. 1 (1976).

[2] M.A. Haase, J. Qiu, J.M. DePuydt, and H. Cheng, Appl. Phys. Lett. 59, 1272 (1991).

[3] W. Richter, Springer Tracts in Modern Physics, Solid State Physics, ed. by G. Höhler, Vol. 78, p. 121; Springer; Berlin, Heidelberg (1976).

[4] M. Cardona, Topics in Applied Physics, Vol. 50, (Light Scattering in Solids II), ed. by M. Cardona and G. Güntherodt, p. 19; Springer; Berlin, Heidelberg (1982).

[5] Semiconductors and Semimetals, Vol. 25: Diluted Magnetic Semiconductors, ed. by J.K. Furdyna, and J. Kossut, Academic Press, 1988.

[6] A.K. Ramdas and S. Rodriguez, in ref. [5], p. 345.

[7] A.K. Ramdas and S. Rodriguez, Topics in Applied Physics, Vol. 68, (Light Scattering in Solids VI), ed. by M. Cardona and G. Güntherodt, p. 137; Springer; Berlin, Heidelberg (1991).

[8] W. Richter, R. Zeyher, and M., Cardona, Phys. Rev. B 18, 4312 (1978).

[9] W. Kauschke, N. Mestres, and M. Cardona, Phys. Rev. B 36, 7469 (1987).

[10] J. Menèndez, L. Viña, M. Cardona, E. Anastassakis, Phys. Rev. B 32, 3966 (1985).

[11] R. Loudon, Adv. Phys. 13, 423 (1964).

[12] M. Rösch, R. Atzmüller, G. Schaack, submitted to Phys. Rev. B (1994).

[13] W. Limmer, Ph.D. Thesis, Univ. Regensburg (1988).

[14] W. Richter, J. Phys. Chem. Solids 33, 2123 (1972).

[15] W. Pötz, P. Vogl, Phys. Rev. B 24, 2025 (1981).

[16] D.L. Peterson, D.U. Bartholomew, U. Debska, A.K. Ramdas, S. Rodriguez, Phys. Rev. B 32, 323 (1985).

[17] W. Hayes, R. Loudon, Scattering of Light by Crystals, J. Wiley & Sons, New York, Chichester, Brisbane, Toronto (1978).

[18] M. Hirsch, R. Meyer, and A. Waag, Phys. Rev. B 48, 5217(1993).

[19] P.A. Wolff, Theory of Bound Magnetic Polarons in Semimagnetic Semiconductors, in ref. [5], p. 413.

[20] T. Dietl, J. Spalek, Phys. Rev. B 28, 1548 (1983).

[21] S.I. Gubarev, T. Ruf, and M. Cardona, Phys. Rev. B 43, 14564 (1991).

[22] Z.C. Feng, A. Mascarenhas, W.J. Choyke, J. Lumin. 35, 329 (1986).

[23] Z.C. Feng, S. Perkowitz, J.M. Wrobel, and J.J. Dubowski, Phys. Rev. B 39, 12997 (1989).

[24] W.J. Keeler, H. Huang, and J.J. Dubowski, Phys. Rev. B 42, 11355 (1990).

[25] T. Daddato, M. Hirsch, G. Schaack, and A. Waag, Proc. XXIIth Int. Conf. on Raman Spectroscopy, ed by J.R. Durig and J.F. Sullivan, p. 440, J. Wiley and Sons, Chichester, (1990).

[26] J. Lambe, C. Kikuchi, Phys. Rev. 119, 1256 (1960).

[27] G.F. Koster, J.O. Dimmock, R.G. Wheeler, H. Statz: Properties of the thirty-two Point Groups, M.I.T. Press, Cambridge, Mass. 1963.

[28] A. Petrou, D.L. Peterson, S. Venugopalan, R.R. Galazka, A.K. Ramdas, and S. Rodriguez, Phys. Rev. B 27, 3471 (1983).

[29] J.A. Gaj, J. Ginter, and R.R. Galazka, phys. stat. sol. (b) 89, 655 (1978).

[30] S. Nagata, R.R. Galazka, D.P. Mullin, H. Akbarzadeh, G.D. Khattak, J.K. Furdyna, and P.H. Keesom, Phys. Rev. B 22, 3331 (1980).

[31] R.R. Galazka, S. Nagata, and P.H. Keesom, Phys. Rev. B. 22, 3344 (1980).

[32] S. Bauer, H. Berger, P. Link, W. Gebhardt, J. Appl. Phys. 74,3916 (1993).

[33] M. Kling, J. Kraus, to be published.

[34] G. Gonzáles de la Cruz, G. Contreras-Puente, F.L. Castillo-Alvarado, C. Mejía-Garcí, and A. Compaan, Solid State Commun. 82, 927 (1992).

[35] A. Ingak, M. L. Bansal, and A. P. Roy, Phys. Rev. B40, 12353 (1989).

On the Microscopic Structures of three Arsenic Antisite-related Defects in Gallium Arsenide studied by Optically Detected Electron Nuclear Double Resonance

J.-M. Spaeth and K. Krambrock

University of Paderborn, Fachbereich Physik,
Warburger Str. 100A, D-33095 Paderborn, Germany

Summary: Three paramagnetic arsenic antisite-related defects in GaAs, which were produced by low temperature electron irradiation and subsequent annealing, have been studied by optically detected electron paramagnetic resonance (ODEPR) and optically detected electron nuclear double resonance (ODENDOR) using the magnetic circular dichroism of the optical absorption (MCDA). Its structure models have been proposed: the isolated As antisite defect, the next nearest anti-structure pair and the arsenic antisite-arsenic interstitial pair defect. The latter defect is the EL2 defect, which is also found in semi-insulating undoped GaAs. The three As antisite-related defects have the feature that their electron paramagnetic resonance (EPR) spectra are identical with the exception of small differences in the inhomogeneously broadened line widths and that they have similar metastable properties upon illumination at low temperatures. Since there is still an ongoing controversy about the microscopic structure of the EL2 defect and since the ODENDOR results were published only in short communications, it is the purpose of this article to show by a detailed comparison of the ODENDOR spectra of the three defects and their analysis what one can conclude from these experiments about the microscopic models and what remains for theoretical interpretation.

Introduction

The EL2 defect is the dominant deep donor in undoped GaAs, which is grown under As-rich conditions. It compensates residual acceptors like the extrinsic impurities C and Zn and grown-in defects involving intrinsic acceptors. Apart from the compensating properties there is a strong interest in EL2 because of its other fascinating properties such as its metastability. The microscopic structure of this defect is still a matter of controversy (for a recent discussion, see ref. [1]). It is clear that this defect is connected with an As antisite, As_{Ga}. At present, mainly two microscopic models are discussed. One is the isolated As_{Ga} defect, which was proposed on the basis of piezo-spectroscopic studies of optical transitions of EL2 [2,3] and favoured by theory to explain the metastable properties [4-6]. The other model is an arsenic antisite-arsenic interstitial (As_{Ga}-As_i) pair proposed first by a combination of EPR and deep level transient spectroscopy (DLTS) in undoped and electron-irradiated n-type GaAs [7] and supported further by optically detected electron nuclear double resonance (ODENDOR) [8,9]. Recently with ODENDOR two different As antisite-related defects were identified in electron-irradiated GaAs. One defect can be produced in low temperature electron-irradiated semi-insulating (SI) GaAs kept at 77 K, and was interpreted as being the isolated As_{Ga} [10], and one is formed in all types of GaAs by electron irradiation at room temperature or by annealing of the low temperature irradiated GaAs crystals near room temperature and was interpreted as being the next nearest anti-structure pair (As_{Ga}-Ga_{As} (nnn))[11]. Similarly as the EL2 defect both antisite-related defects show a light-induced metastability at low temperature. They recover back to the ground state by heating the samples to 140 K.

Detailed structure information about defects can be obtained by electron paramagnetic resonance (EPR) provided the hyperfine (hf) and superhyperfine (shf) interactions can be resolved. However, for intrinsic defects in GaAs, and especially for the three As antisite-related defects, only the hf interaction with the central nucleus is resolved by EPR, which results in the typical four-line hyperfine split quadruplet spectrum without further splittings preventing a detailed microscopic structure identification (the nuclear spin of ^{75}As is $I = 3/2$ with 100 % abundance). The particular difficulty is that the hf split EPR spectra are almost identical: within experimental error they have the same hf splitting. The spectra differ only slightly in the line widths of the individual hf lines. To resolve the interactions with neighbouring nuclei electron nuclear double resonance (ENDOR) measurements are necessary. Conventional ENDOR measurements in GaAs have failed because of the rather low signal-to-noise ratio of the EPR signals. In recent years it could be shown that the optical detection of EPR (ODEPR) and ENDOR (ODENDOR) via the magnetic circular dichroism of the optical absorption (MCDA) has enhanced sensitivity and was able to resolve shf interactions from neighbouring nuclei thus providing more information about the microscopic structure of these three intrinsic defects. In view of the ongoing controversy about the microscopic structure of the EL2 defect, it seems desirable and important to compare and to summarize the ODENDOR investigations of the three different As antisite-related defects, which have been published only in relatively short separate communications [8, 10, 11]. Unfortunately, ENDOR spectroscopy is somewhat complicated and in difficult cases it is not always easy to see which are the conclusions as to microscopic defect models to be drawn from the spectra. A comparison and more detailed description for the three As_{Ga}-related defects should therefore be helpful to see that inspite of the "identical" EPR spectra and inspite of very similar metastability properties the three defects have different microscopic structures. In particular for the EL2 defect we show that the pair model As_{Ga}-As_i is not the only possible structure which would be compatible with the ENDOR spectra.

More As_{Ga}-related defects have been produced in GaAs, although in as-grown GaAs only EL2 appears. The other defects differ in hf interaction, spin-lattice relaxation time and optical properties from those discussed above. No ENDOR investigations are available yet. Therefore, comparison with the properties of the defect having been assigned to the isolated As_{Ga} defect leads to the conclusion that they must all be As_{Ga}-complexes, which contain the As_{Ga} as one constituent.

1 ODENDOR results of three arsenic antisite defects in GaAs

1.1 Experimental method and samples

The preparation of the samples was described elsewhere [10,11]. The method of optical detection of EPR and ENDOR is based on the measurement of the magnetic circular dichroism of the optical absorption (MCDA) in a longitudinal magnetic field. Optically detected EPR is measured as a microwave-induced change of the MCDA. Optically detected ENDOR is recorded as a radio-frequency-induced change of the MCDA under microwave resonance (EPR) conditions. For further details of the spectrometer and the experimental method see ref. [12]. Only one point should be emphasized. Apart from the sensitivity enhancement in the optical detection compared to conventional EPR/ENDOR it is possible to correlate the EPR/ENDOR information (e.g. the hf interaction) with the optical absorption spectra of the defect by measuring the so-called "MCDA tagged by EPR/ENDOR", a kind of ODEPR or ODENDOR excitation spectra [12]. Thus, the EPR or ENDOR spectra of a defect are correlated with its MCDA spectra. This is of particular importance if e.g. the hf interaction of two defects is the same, although their structures and their optical spectra are different, as is the case for several As_{Ga} defects.

1.2 Analysis of the (OD)ENDOR spectra

With ENDOR spectroscopy it is possible to determine the shf and quadrupole interactions, which are hidden in an inhomogeneously broadened EPR line (for a recent comprehensive description of the method see [12]). The energy of a spin system which consists of an unpaired electron with spin $\boldsymbol{S}$, a central nucleus with nuclear spin $\boldsymbol{I}_c$ and N neighbours with nuclear spins $\boldsymbol{I}_j$ is described by the spin Hamiltonian $\boldsymbol{H}$ [13]:

$$\boldsymbol{H} = \beta_e \boldsymbol{B} g \boldsymbol{S} + \boldsymbol{I}_c \boldsymbol{A}_c \boldsymbol{S} - g_{I,c}\beta_n \boldsymbol{B}\boldsymbol{I}_c + \boldsymbol{I}_c \boldsymbol{Q}_c \boldsymbol{I}_c + \sum_{j=1}^{N} (\boldsymbol{S}\boldsymbol{A}_j \boldsymbol{I}_j + \boldsymbol{I}_j \boldsymbol{Q}_j \boldsymbol{I}_j - g_{I,j}\beta_n \boldsymbol{B}\boldsymbol{I}_j) \tag{1}$$

The first term is the electron Zeeman term, the next three terms describe the hf interaction, the nuclear Zeeman term and the quadrupole interactions of the unpaired electron with the central nucleus and the other terms in the sum are the shf and quadrupole interactions with all further interacting N neighbouring nuclei. The symbols have their usual meaning: β_e is the Bohr magneton and β_n is the nuclear magneton, $\boldsymbol{S}$ and $\boldsymbol{I}_j$ are the electron and nuclear spin operators, respectively, $\boldsymbol{B}$ is the static magnetic field, g is the g-tensor, g_I is the nuclear g-factor and $\boldsymbol{A}_c$ and $\boldsymbol{Q}_c$ are the hf and quadrupole tensors of the central nucleus, and $\boldsymbol{A}_j$ and $\boldsymbol{Q}_j$ are the shf and quadrupole tensors of the interacting nuclei j. For ENDOR transitions the selection rules are: $\Delta m_s = 0$ and $\Delta m_I = \pm 1$.

In their principal axis system the hf and shf tensors $\boldsymbol{A}_c$ and $\boldsymbol{A}_j$, respectively, are often decomposed into an isotropic and an anisotropic part according to

$$\boldsymbol{A} = a\mathbf{1} + \boldsymbol{B} \tag{2}$$

$$b = \tfrac{1}{2}B_{zz} \tag{3}$$

$$b' = \tfrac{1}{2}(B_{xx} - B_{yy}) \tag{4}$$

x, y, z is the principal axis system of the hf or shf tensors, a is the isotropic hf or shf constant (Fermi contact term), b is the anisotropic hf or shf constant and b' describes the deviation of the tensor from axial symmetry. The largest interaction is along the z direction. The quadrupole tensor is traceless and contains the quadrupole interaction constants q and q':

$$q = \tfrac{1}{2}Q_{zz} \tag{5}$$

$$q' = \tfrac{1}{2}(Q_{xx} - Q_{yy})\,. \tag{6}$$

For the sake of simplicity we have omitted the subscripts c or j, which would indicate that each nucleus interacting has his own set of interaction constants and tensors, respectively.

In order to discuss the analysis of the ENDOR spectra we present a solution of equation (1) in perturbation theory of first order, but it should be noted already at this point that this solution is inadequate for the exact description of the ODENDOR spectra of the arsenic antisite-related defects in GaAs. Equation (1) has to be solved by numerical diagonalization.

For the analysis of the ENDOR transitions equation (1) can be reduced to the "ENDOR Hamiltonian" which is given by the term with the sum over the neighbouring nuclei with the assumption that the shf and quadrupole interactions are small compared to the electron Zeeman term (first term in equation (1)) and to the interactions with the central nucleus. In first order the quantization of the electron spin is not affected by these interactions and the nuclei are independent of each other. They can be treated separately and the sum in equation (1) can be omitted [12].

In order to solve the ENDOR Hamiltonian in perturbation theory of first order it is further assumed that the anisotropic shf interactions (b_j, b'_j) and the quadrupole interactions (q_j, q'_j) are small compared to the isotropic shf interactions a_j. The calculation of the energy eigenvalues of the ENDOR Hamiltonian for one neighbouring nucleus then yields:

$$E = W_{shf} m_S m_I - g_I \beta_n B m_I + \tfrac{1}{2}(m_I^2 - \tfrac{1}{3} I(I+1)) W_q \quad (7)$$

with the following abbreviations:

$$W_{shf} = a + b(3\cos^2\Theta - 1) + b'\sin^2\theta\cos 2\delta \quad (8)$$

$$W_q = 3q(3\cos^2\Theta' - 1) + q'\sin^2\Theta'\cos 2\delta' \quad (9)$$

where Θ, δ and Θ', δ' are the polar angles of $\boldsymbol{B}$ in the principal shf and quadrupole axes systems, respectively.

According to the selection rule for the nuclear magnetic resonance transitions, the ENDOR frequencies are:

$$\nu_{ENDOR} = \frac{1}{h}\left|m_S W_{shf} + m_q W_q - h\nu_n\right| \quad (10)$$

with

$$\nu_n = \frac{g_I \beta_n B}{h}\,. \quad (11)$$

ν_n is the Larmor frequency of a free nucleus in the magnetic field B and $m_q = 1/2(m_I + m_{I'})$, where m_I and $m_{I'}$ are the two nuclear spin quantum numbers connected by the ENDOR transition.

From the measurement of the magnetic field dependence of the ENDOR frequencies it is generally possible to determine the chemical identity of the interacting nuclei. According to equations (10, 11) the ENDOR frequency contains the Larmor frequency which is directly proportional to the magnetic field. From a measurement of the field dependence of the ENDOR line frequencies one can determine the characteristic g_I factor. This "field shift" method can also be applied when the ENDOR Hamiltonian has to be diagonalised [12]. In first order, according to equation (10), for $S = 1/2$ and without quadrupole interaction, each nucleus gives a pair of lines because of $m_s = \pm 1/2$ separated by $2\nu_n$ if $1/2W_{shf} < h\nu_n$, and separated by W_{shf}, if $h\nu_n > W_{shf}$. Therefore, if the magnetic field is known, the nuclei can be identified by their g_I-factors either from the line pairs separated by $2\nu_n$ or by symmetric line patterns about ν_n.

In order to determine the interaction parameters and the symmetries of the interaction tensors, the dependence of the ENDOR frequencies upon the variation of the magnetic field with respect to the crystal orientation must be measured and analysed. For the calculation of an ENDOR angular dependence it is necessary to assume a possible defect model. The calculated angular dependence, for example with a diagonalization of equation (1), is then compared to the measured one. An agreement is only achieved for the correct defect model [12].

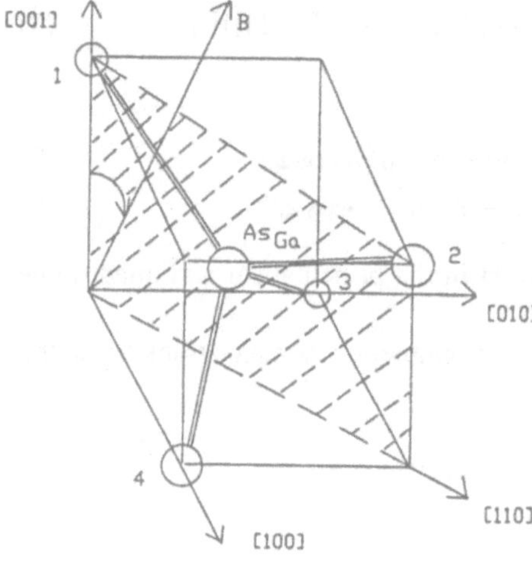

Figure 1
The As antisite defect with its first four tetrahedrally symmetric As neighbours (1-4) and the crystal orientations with respect to the magnetic field. The rotation plane of the magnetic field is a $\{110\}$ plane. The rotation angle 0 corresponds to the [001] crystal direction. The neighbours 3 and 4 are equivalent for the whole angular dependence.

Now we discuss what kind of ENDOR spectrum and angular dependence we expect for the nearest neighbours of the isolated arsenic antisite defect (henceforth called simply "antisite"). Figure 1 shows the defect with its four nearest As neighbours in $<111>$ directions. For the measurement of the angular dependence the magnetic field is rotated in a $\{110\}$ plane. All nuclei which have the same distance from the antisite and which can be transformed into each other by symmetry operations of the crystal are said to belong to one "shell".

The interaction tensors $\boldsymbol{A}$ and $\boldsymbol{Q}$ are symmetric second rank tensors. Thus, there exist at most six independent interaction parameters for each tensor. In the principal axis system these are the three principal values and three Euler angles Θ, Ψ, and Φ to describe the orientation of the principal axis system in the crystal. Each nucleus has its own principal axis system for its shf and also for its quadrupole tensor. Often the tensor orientation in the crystal is determined by symmetry. If the defect centre and the respective nucleus are in a mirror plane of the defect, then two principal axes must be in the mirror plane. If the connecting line between the neighbouring nucleus and the centre is a threefold or higher symmetry axis, then the tensor is axially symmetric with its axis in the symmetry axis. For the nearest neighbours of the isolated antisite defect one expects that the parameters b' and q' are zero and that there is no free Euler angle, since the $<111>$ directions are threefold symmetry axes.

For a "$<110>$" symmetry (which the second and third shell neighbours of the antisite have) the lattice is invariant under a reflection at a $\{110\}$ plane. In this case the parameters b' and q' are not zero, and there is one free Euler angle Θ for the tensor orientation. The next neighbour shells of an arsenic antisite defect with its symmetries, the number of nuclei and its distances from the antisite are shown in table 1.

The expected angular dependence of the ENDOR lines of the four nearest ^{75}As neighbours of the isolated As_{Ga} according to equation (10) is illustrated in figure 2(a) for $a/h = 150$ MHz, $b/h = 50$ MHz and $m_S = \pm 1/2$ and in figure 2(b) with

Table 1 Distances and symmetries of the neighbour shells of an As antisite defect in GaAs

no.	atom	symmetry	number	distance [Å]
1	As	<111>	4	2.44
2	Ga	<110>	12	3.95
3	As	<110>	12	4.67
4	Ga	<100>	6	5.61
5	As	<110>	12	6.13
6	Ga	<110>	24	6.85
7a	As	<110>	12	7.37
7b	As	<111>	4	7.37

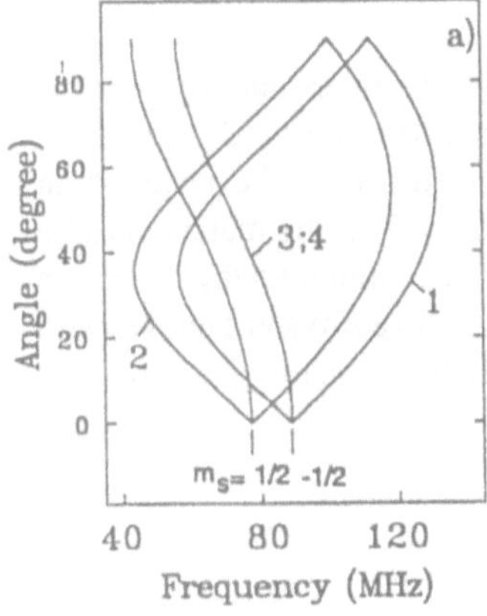

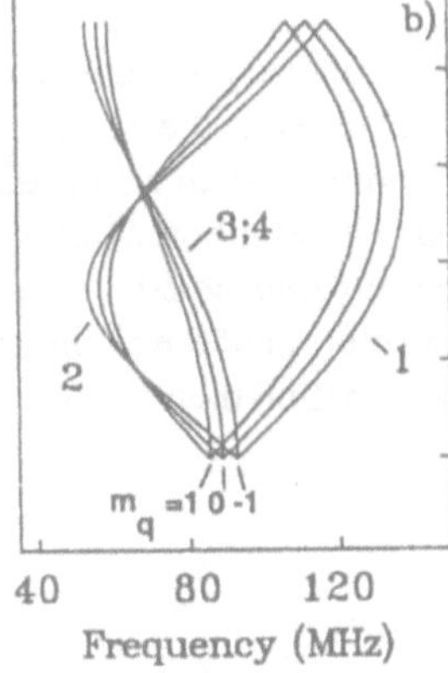

Figure 2
Calculated angular dependence of the four nearest neighbours of the As antisite defect (a) with $a/h = 150$ MHz and $b/h = 50$ MHz for both m_S states and (b) with an additional quadrupole interaction of $q/h = 1$ MHz, only for $m_s = -1/2$. For details see text.

an additional quadrupole interaction of $q/h = 1$ MHz, only for $m_S = -1/2$ (the lines for $m_S = +1/2$ are omitted). Since $I(^{75}\text{As}) = 3/2$, because of the quadrupole interaction the "shf" lines of figure 2(a) split into triplets, since $m_q = -1, 0, +1$ in equation (10). The curvature of the angular dependence is now determined by both b and q. As long as q remains small, figure 2(b) looks somewhat similar to figure 2(a). In the $\{110\}$ rotation plane two of the four nuclei (nuclei 3 and 4, see figure 1) remain equivalent with respect to the magnetic field B, therefore they have the same angular dependence. (Their tensor orientations with respect to B are always the same for B in the (110) plane). The angular dependencies shown in figure 2 are characteristic for the "<111>" symmetry of the nearest neighbours of the antisite defect. The picture changes drastically for a larger quadrupole interaction (see e.g. below). In this case the assumption $b, q << a$ made to derive equation (10) is no longer valid. The calculation of the angular dependence must be done by numerical diagonalization of the ENDOR Hamiltonian.

Because of the large central hf interaction of the antisite-related defects (a_c/h is nearly 25 % of the electron Zeeman interaction for microwave frequencies of 24 GHz), the electron spin is not exactly $S = 1/2$. An approximation to deal with this is the "effective spin" concept. In order to calculate the ENDOR frequencies, the electron Zeeman term and the central hf term are diagonalized numerically first. With the result of this diagonalization the effective spin quantum number $m_{S\,eff}$ is calculated as the expectation value of the z component of the electron spin. The effective spin quantum number $m_{S\,eff}$ is then used instead of m_S in the ENDOR Hamiltonian.

For the first shell 18 ENDOR lines from the four nearest As neighbours (nuclei 3 and 4 are equivalent) are expected because of two m_s and three m_q states (9 lines are shown in figure 2(b), for $m_s = -1/2$). This is valid as long as the 4 neighbour nuclei can be treated separately, i.e. without pseudo-nuclear coupling (see below).

More ENDOR lines are measured from the four nearest neighbours either because of symmetry lowering, i.e. if by some perturbation the four nearest neighbours do not have exactly the same shf and quadrupole tensors any more or because of the effect of pseudo-dipolar coupling [14-16].

So far in the ENDOR Hamiltonian each nucleus was treated independently. For the exact calculation of the problem, however, the nuclei cannot be treated independently, the full spin Hamiltonian (equation (1)) must be diagonalized. In the case of an arsenic antisite with its four nearest neighbours, all with $I = 3/2$, the matrices to be diagonalized reach a dimension of (2048 × 2048). With our computers this problem could not be calculated so far. A solution of the problem for two equivalent nuclei with $I = 3/2$ (nuclei 3 and 4) by perturbation theory is given in the literature [14-16].

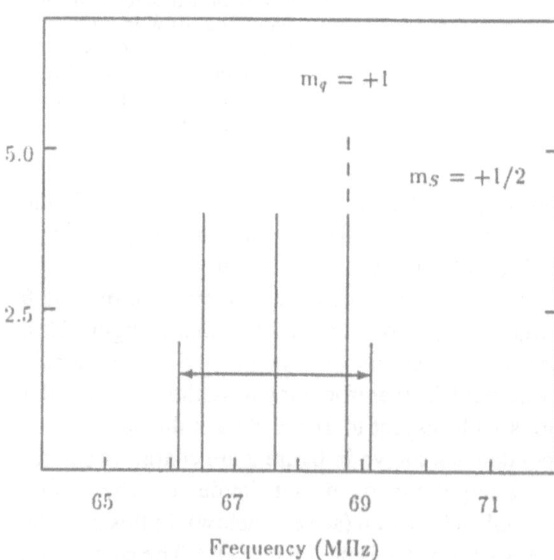

Figure 3
Calculated splitting of the ENDOR line with $m_s = 1/2$, $m_q = 1$, $B = 650$ mT parallel to [001], coming from the equivalent nuclei 3 and 4 because of the effects of pseudo-dipolar coupling using the approximation given in [15] ($a/h = 150$ MHz, $b/h = 50$ MHz). The overall splitting is 3.04 MHz. The position of the ENDOR line without consideration of the pseudo-dipolar coupling is indicated in the figure by the dashed line.

Following the perturbation solution for two equivalent nuclei [15], the ODENDOR lines split for $m_q = 0$ into four and for $m_q = \pm 1$ into five lines. Figure 3 shows the line splittings of the line of the equivalent nuclei 3 and 4 with $m_q = 1, m_s = 1/2$ according to this perturbation solution. The overall splitting is 3.04 MHz for $B = 0.65$ T and is 2.08 MHz for $B = 0.95$ T ($B||[001]$ and assuming as before $a/h = 150$ MHz, $b/h = 50$ MHz).

For the As antisites such a perturbation solution is not sufficient to describe the whole angular dependence of the four nearest neighbours. The non-equivalent nuclei 1 and 2 cannot be described within this perturbation solution. At present the results of the numerical diagonalization of the full spin Hamiltonian (equation (1)) for the central atom (As_{Ga}) and for two of the four nearest neighbours (matrices of dimensions of (128×128)) are available (see next section).

One point of the ENDOR analysis was not mentioned so far. From an ENDOR analysis neglecting pseudo-dipolar coupling effects it is not possible to decide whether the measured angular dependence of the first As neighbour shell comes from four As nuclei or whether one or several As nuclei are absent, as long as the geometry remains the same. If, for example, one nucleus was missing, then there would be four centre orientations in the crystal and the ENDOR lines of all four nuclear positions would appear as if they were all occupied by a nucleus. A decision on the number of neighbour nuclei may be possible if the pseudo-dipolar coupling can be calculated accurately for all neighbours. At present this is not possible.

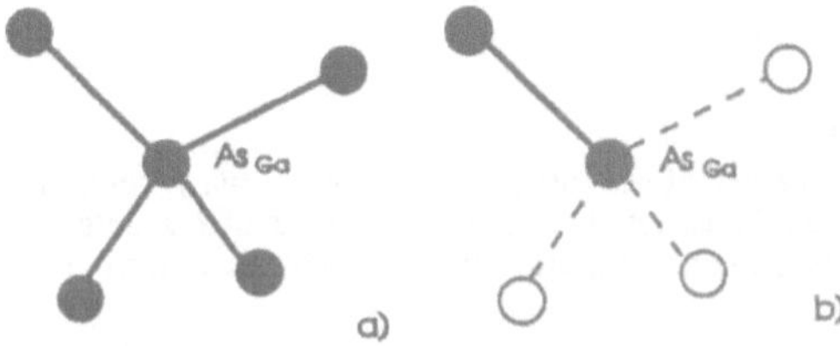

Figure 4
(a) An As antisite defect with its four nearest neighbours and (b) with only one neighbour. As long as the geometry remains the same both cases cannot be distinguished by ENDOR if the effect of pseudo-dipolar coupling can be neglected.

Figure 4 illustrates the case where the antisite is surrounded by its four nearest neighbours (a) and where the antisite is surrounded only by one neighbour (b). In case (b), if the geometry remains the same as in case (a), without pseudo-dipolar coupling the ENDOR angular dependence would be the same as in case (a), because there exist four centre orientations for this neighbour. One must therefore calculate the line shape (width) of the EPR spectrum in order to decide how many neighbour nuclei are present. For this calculation the shf constants determined from the ENDOR analysis are used. In the case (a) the EPR line consists of 13 shf lines ($2 \cdot I \cdot N + 1 = 2 \cdot 3/2 \cdot 4 + 1 = 13$) split by the shf interaction with the intensity ratio $1 : 4 : 10 : 20 : 31 : 40 : 44 : 40 : 31 : 20 : 10 : 4 : 1$, while in case (b) only 4 lines ($N = 1$) with the same splitting between each line and equal intensities are expected, i.e. a much narrower EPR line with different shape. Thus, the calculation of the EPR line shape with the shf constants determined from ENDOR represents a consistency check for the ENDOR analysis (see also ref. [12] on this point).

1.3 Experimental results

EPR investigations of GaAs crystals grown under different melt stoichiometries have shown that the EL2 defect is an arsenic antisite-related defect [17,18]. The EPR spectrum and likewise the ODEPR spectrum (figure 5(b)) of the EL2 defect in its paramagnetic charge state ($EL2^+$) are characterized by an isotropic quadruplet due to the hf interaction of an unpaired electron with the central ^{75}As nucleus. Shf interactions are not resolved, the four lines are inhomogeneously broadened. No further structural details are thus discernable in the EPR spectrum.

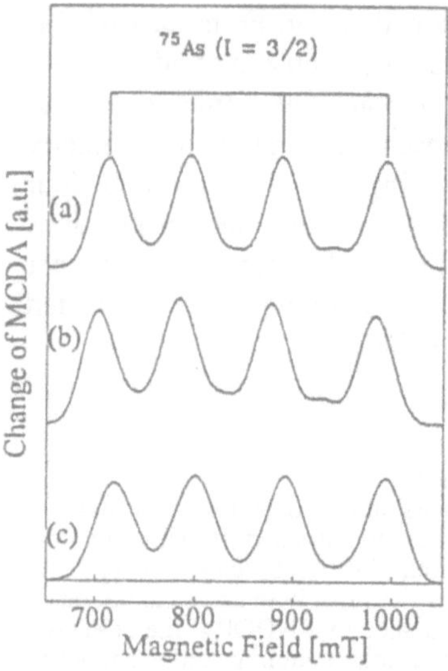

Figure 5
ODEPR spectra of the three different As antisite-related defects: (a) the isolated As_{Ga} defect, (b) the EL2 defect and (c) the next nearest anti-structure pair measured at 1.5 K.

Recently it was shown that two other As antisite-related defects with nearly the same ODEPR spectra could be produced by electron irradiation. The ODEPR spectrum of the "isolated" As antisite defect, As_{Ga}^+, (figure 5(a)) can be observed after electron irradiation of SI GaAs at low temperature (4.2 K), when the sample is kept below and up to 77 K [10], that of the next nearest anti-structure pair (As_{Ga}^+-Ga_{As} (nnn)) after electron irradiation of all types of GaAs at low temperature and subsequent annealing to near room temperature (RT) or directly after electron irradiation at room temperature [11] (figure 5(c)). The three ODEPR spectra are nearly identical with the exception of small differences in the inhomogeneously broadened line widths. In particular the hf splitting of the quadruplet lines is practically identical for all three defects. The EPR parameters are listed in table 2.

From the ODEPR excitation spectra (MCDA tagged by EPR) measured of the different As antisite-related defects it is clear that they must come from different microscopic structures, which, however, do not show up in the ODEPR spectra. Figure 6 shows the ODEPR excitation spectra of (a) the isolated As antisite defect, (b) the $EL2^+$ defect and (c) the next nearest anti-structure pair. They are all clearly different

Table 2 ODEPR parameters of three different As antisite-related defects in GaAs: a/h is the isotropic hyperfine constant of the central ^{75}As nucleus, g is the electronic g factor and $\Delta B_{1/2}$ is the halfwidth of each of the hf split ODEPR lines.

	a/h[MHz] for ^{75}As	g	$\Delta B_{1/2}[mT]$
isolated As_{Ga}	2650	2.04	32 ± 1
EL2	2656	2.04	34 ± 1
anti-structure pair	2600	2.04	40 ± 1

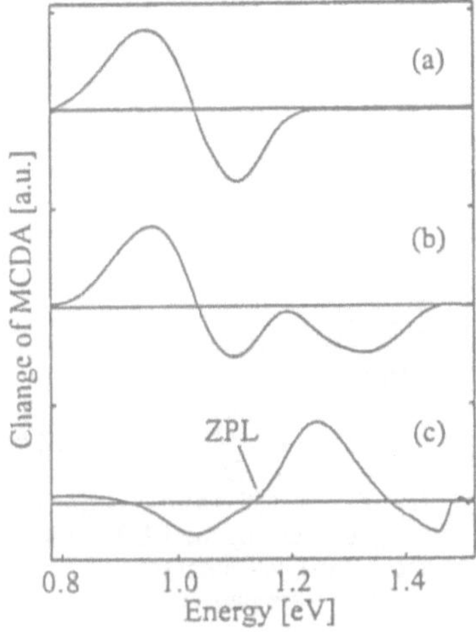

Figure 6
ODEPR excitation spectra (MCDA tagged by EPR) of the three different As antisite-related defects: (a) the isolated As_{Ga} defect, (b) the EL2 defect and (c) the next nearest anti-structure pair. In (c) the so-called zero phonon line of the anti-structure pair is indicated [11].

from each other. For the determination of the microscopic structures of these different As antisite-related defects ODENDOR measurements were performed in order to resolve the shf interactions with the neighbouring lattice nuclei. In the following section the ODENDOR spectra of the three antisite-related defects are compared to demonstrate that it is also seen there that the three defects do have different microscopic structures.

ODENDOR of the nearest neighbours

Typical ODENDOR spectra of the nearest neighbours of the three As antisite-related defects in the frequency range from 35 to 120 MHz are shown in figure 7 (a-c) for $B||[001] + 18°$ in a $\{110\}$ plane: fig. 7(a) the isolated As_{Ga} defect, fig. 7(b) the EL2 defect and fig. 7(c) the next nearest anti-structure pair. The ODENDOR spectra of the isolated As_{Ga} defect and of the EL2 defect are nearly identical in this frequency range except for a few weak lines marked by As_i (see below). The spectrum of the

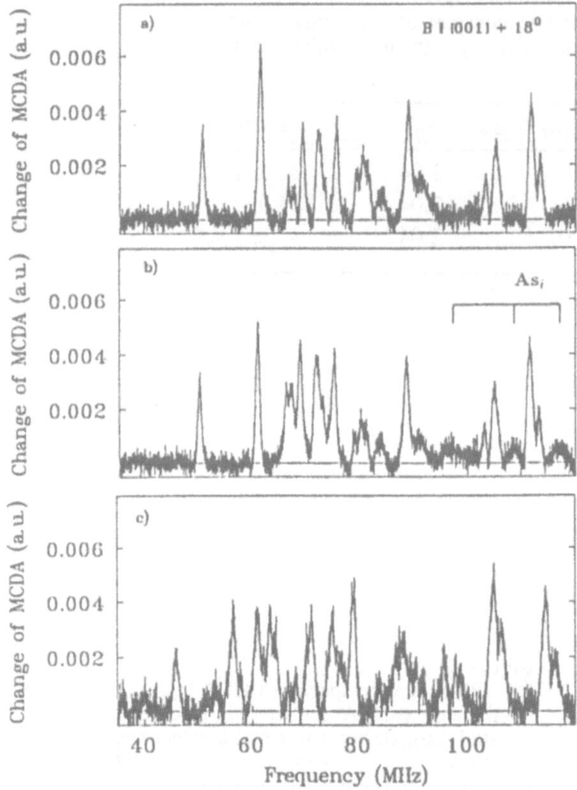

Figure 7
ODENDOR spectra of three different As antisite-related defects in the frequency range from 35 to 120 MHz measured in the low field ODEPR lines for $B||[001] + 18°$: (a) the isolated As_{Ga} defect, (b) the EL2 defect and (c) the next nearest anti-structure pair.

anti-structure pair is clearly different from the two other ones.

The ODENDOR effect depends strongly on the field position within the ODEPR line at which it was measured. It is largest in the flanks of the ODEPR lines. This is true for all three As antisite-related defects discussed here. In addition, the ODENDOR effect is of the same order of intensity as the ODEPR effect, which has never been observed in conventional stationary ENDOR experiments of a solid state defect. A qualitative explanation could be an effect like the dynamical nuclear-nuclear spin polarisation [13], but it is not understood in detail. ODENDOR experiments of the phosphorus antisite defects in InP [19] and GaP [20] showed that the ODENDOR effect is of a similar magnitude, but it is largest in the maximum of the ODEPR lines.

First the chemical identity of the interacting nuclei giving rise to the lines must be determined. Figure 8(a) shows the ODENDOR lines of the isolated As_{Ga} defect in the frequency range from 45 MHz to 70 MHz for different magnetic field values and $B||[001] + 18°$. The marked ODENDOR lines belong to the nucleus no. 2 of figure 1 of the first As shell with $m_q = 0$ and $m_s = \pm 1/2$. According to equations (10, 11) the positions of the ODENDOR lines depend on the magnetic field. From the slope

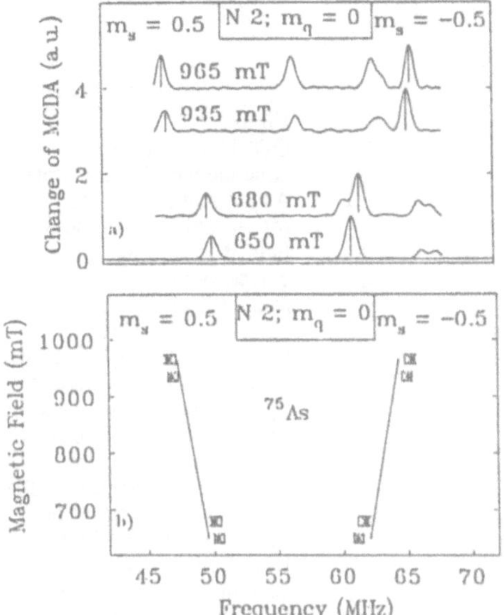

Figure 8
Magnetic field dependence of the ODENDOR lines in the frequency range from 45 to 70 MHz for $B||[001] + 18°$ (in a $\{110\}$ plane). (a): The marked ODENDOR lines belong to the As neighbour no. 2 with $m_q = 0$ for different magnetic fields. (b): Measured line positions (represented by stars), calculated field shift assuming a linear dependence (equation (10): solid lines in figure 8(b)) and calculated field positions taking into account the effective spin and using a numerical diagonalisation of the ENDOR Hamiltonian (squares).

of the straight line (figure 8(b)) one can determine the nuclear g_I factor and therefore the chemical element. In first order the two straight lines were calculated according to equation (10) for ^{75}As for the two m_s states. The stars represent the measured line positions. The calculated and measured line positions do not agree. The measured magnetic field dependence of the ENDOR lines is not quite linear.

The observed lines can be explained taking into account the effective spin since the central ^{75}As hf interaction is not small compared to the electron Zeeman term. It could be shown that all ODENDOR lines arise from ^{75}As nuclei. Their calculated positions for the different field values are the squares in figure 8(b), which excellently agree with the experimental positions (stars). Similarly, the ODENDOR lines of the EL2 defect [8,9] and the anti-structure pair in this frequency range were shown to be due to ^{75}As. All lines above 30 MHz belong to ^{75}As nuclei.

The isolated As antisite defect

In order to determine the symmetry and the interaction parameters of the As ligands, spectra like those in figure 7 have to be measured for different angle positions of the crystal with respect to the magnetic field. In the case of the isolated As_{Ga} defect the ODENDOR angular dependence was measured in steps of 1.5° from −30° to +30° for a rotation of the crystal in a $\{110\}$ plane (see in figure 1: 0° corresponds to [001], and 90° to [110] crystal orientation).

The ODENDOR angular dependence of the isolated As_{Ga} defect in the frequency range from 30 to 130 MHz is shown in figure 9(a) measured in the ODEPR line at

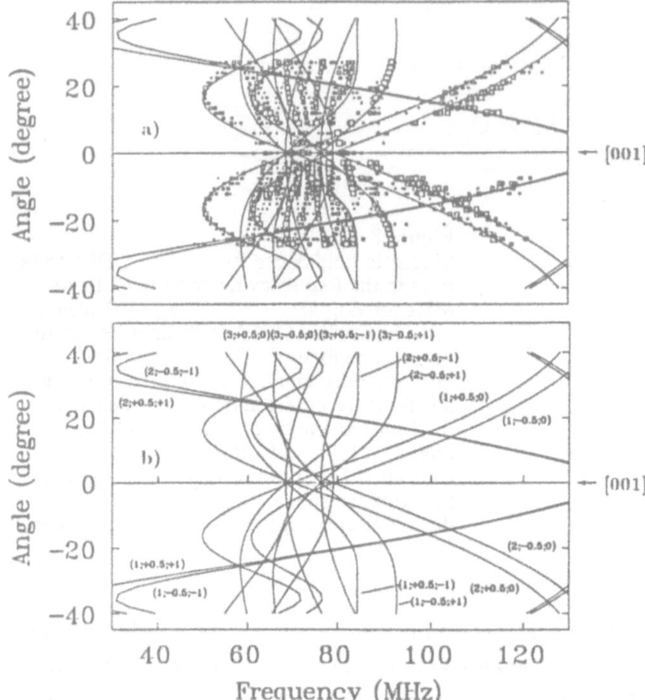

Figure 9
ODENDOR angular dependence of the isolated As antisite defect for rotation of the magnetic field B in a {110} plane. 0° corresponds to B parallel to the [001] crystal direction. (a): The measured ODENDOR line positions are represented by the squares and the solid lines represent the result of the a simulation of the angular dependence for the four nearest As neighbours by solving the ENDOR Hamiltonian taking into account the effective spin in second order. (b): Calculated angular dependence as in (a) with line assignments: The tripels mark the number of the nucleus, the spin state and the m_q state.

the lowest field ($m_I = 3/2$). The squares represent the frequency positions of the ODENDOR lines and their magnitudes are a measure of the line intensities. The calculation of the angular dependence was done with the assumption that these ODENDOR lines belong to the four tetrahedrally coordinated nearest As neighbours of the As antisite with <111> symmetry. For the calculation the ENDOR Hamiltonian was numerically diagonalized, but with the approximation, that each nucleus is independent, e.g. the pseudo-dipolar couplings between the different nuclei are not considered. The parameters for the calculation are: $a/h = 169.3$ MHz, $b/h = 53.2$ MHz and $q/h = 12.0$ MHz (solid lines in figures 9(a) and 9(b)). The effective spin was taken into account in second order. The number tripels in figure 9(b) describe the position of the nucleus (nuclei 1-4; figure 1), the spin state m_s ($m_s = \pm 1/2$) and the quadrupole transition m_q ($m_q = 1, 0, -1$). The curvature of the angular dependence is determined by the large quadrupole interaction (compare figure 2). Only for $m_q = 0$ the two m_s curves are nearly parallel. In the [001] direction their frequency separation is $2\nu_n$. The anisotropic interaction vanishes for $B||[001]$. The calculated angular dependence describes approximately the experimentally observed angular dependence. However, many observed smaller line splittings are not explained by this calculation.

These line splittings can have different reasons:

1. Pseudo-dipolar couplings, where the nuclear spins of the different nuclei interact with each other through the electron spin, do lead to line splittings.

2. The crystal was not exactly orientated in the (110) plane. Then nuclei 3 and 4 loose their equivalence. However, only the ODENDOR lines of the nuclei 3 and 4 can split by this effect. Therefore, we can exclude such an effect.

3. If the nearest neighbours are not precisely tetrahedrally coordinated, then there could be up to four different shf and quadrupole tensors and one would have the superposition of up to four centre orientations. However, the angular dependence cannot be explained by such an effect (see also the results for the anti-structure pair below).

Therefore, the splittings must be explained by the effect of pseudo-dipolar coupling of the nuclei.

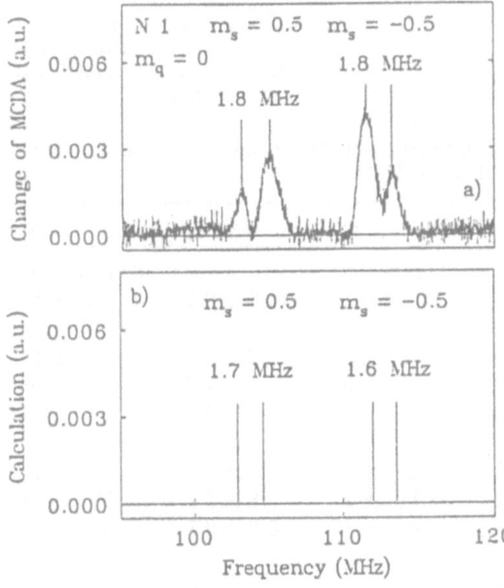

Figure 10
Part of the ODENDOR spectrum of figure 7(a). (a): Observed line splittings of the ODENDOR lines which belong to the single nucleus no. 1 and $m_q = 0$. (b): Calculated ODENDOR line splitting by numerical diagonalisation of the full spin Hamiltonian (equation (1)) for the single nucleus 1.

A part of the ODENDOR spectrum (figure 7(a)) in the frequency range from 95 to 120 MHz is shown in figure 10(a). From the calculation of the ODENDOR angular dependence it follows that the observed lines belong to the single nucleus 1. Two doublets can be seen, where the two lines of the doublets are separated by 1.8 MHz. The lower doublet belongs to the $m_s = +1/2$ and the higher one to the $m_s = -1/2$ state with $m_q = 0$. The splitting is anisotropic. For the [001] direction it is largest with a separation of 2.4 MHz. In this orientation all four nuclei are equivalent with respect to the magnetic field.

In order to calculate the pseudo-dipolar coupling effects the full spin Hamiltonian must be diagonalized. For As antisite-related defects with $S = 1/2$ and $I_{\mathrm{As}} = 3/2$

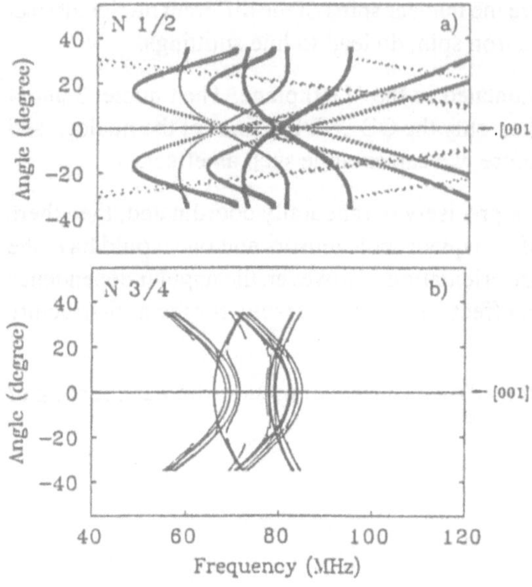

Figure 11
Calculation of the ODENDOR angular dependence taking into account the pseudo-dipolar coupling. (a): The full spin Hamiltonian for the central atom and the single nuclei no. 1 and 2 was diagonlised numerically. (b): The full spin Hamiltonian for the central atom and the equivalent nuclei 3 and 4 was diagonalised numerically.

Table 3 Superhyperfine and quadrupole interaction constants of the isolated As antisite defect "As_{Ga}" and of the EL2 defect (After [8,10] and new calculations.)

	a/h [MHz]	b/h [MHz]	Θ_b	q/h [MHz]	Θ_q
1st As shell			<111> $\hat{=}0°$		<111> $\hat{=}0°$
As_{Ga}	169.3 ± 0.1	53.2 ± 0.1	$0° \pm 1°$	12.0 ± 0.1	$0° \pm 1°$
EL2 nucl. 1	167.8 ± 0.1	53.9 ± 0.1	$0° \pm 1°$	11.9 ± 0.1	$0° \pm 1°$
EL2 nucl. 2-4	169.3 ± 0.1	53.9 ± 0.1	$0° \pm 1°$	11.7 ± 0.1	$0° \pm 1°$
EL2: As_i	215	44	0°	4.8	0°
	232	13	30°	1.4	30°
2nd As shell			<110> $\hat{=}0°$		<110> $\hat{=}0°$
As_{Ga}: As_{III}	21.5 ± 0.1	2.2 ± 0.1	$28° \pm 2°$	0.5 ± 0.1	$30° \pm 1°$
EL2: As_{IIIa}	$\pm 35.2 \pm 0.1$	$\mp 1.3 \pm 0.1$	$11° \pm 1°$	2.8 ± 0.1	$25° \pm 1°$
EL2: $As_{IIIb,c}$	19.5 ± 0.1	3.2 ± 0.1	$15° \pm 1°$	0.9 ± 0.1	$32° \pm 1°$

and four neighbours matrices of the dimension of 2×4^5, i.e. (2048×2048) must be diagonalized. This was not possible, so far. However, calculations considering only two neighbour nuclei and the central nucleus (matrices of dimensions (128×128)) show that the calculated and measured line splittings are of the same order and that such a splitting occurs also for the single (non-equivalent) nuclei no. 1 and 2. The results of this calculation for $a/h = 169.3$ MHz, $b/h = 53.2$ MHz and $q/h = 12.0$ MHz are shown in figure 10(b) for nucleus 1 and the corresponding angular dependence is shown in figure 11(a) for the single nuclei 1 and 2 and in figure 11(b) for the equivalent nuclei 3 and 4. Both angular dependencies together agree well with the measured ODENDOR angular dependence. Therefore, the observed line splittings of the ODENDOR lines of the isolated As antisite defect are explained by the effect of pseudo-dipolar coupling. The shf and quadrupole parameters are listed in table 3.

As a further example for such ENDOR line splittings those observed for the Ga vacancy in GaP [16] are referred to. The shf constants ($a/h = 195$ MHz and $b/h = 54$ MHz) are of the same order as for the As antisite-related defects. In that case the observed line splittings could also be explained by the effect of pseudo-dipolar coupling. Experimentally this was verified by double ENDOR measurements [16]. For the calculation of the angular dependence the spin Hamiltonian was numerically diagonalized taking into account the electron spin and the nuclear spins of all four neighbouring nuclei. In the case of the Ga vacancy ($S = 3/2$ and $I_P = 1/2$) only matrices of dimensions of (64×64) had to be diagonalised. The calculation could explain the observed splittings very well [16].

The EL2 defect

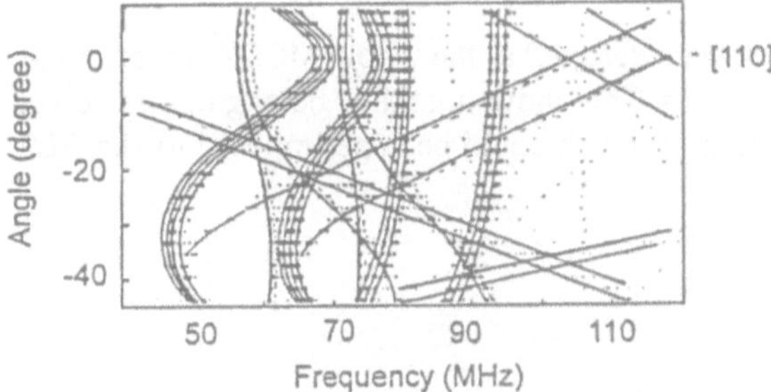

Figure 12
ODENDOR angular dependence of the nearest neighbours of the EL2 defect, after [9].

Figure 12 shows the ENDOR angular dependence of the nearest ^{75}As neighbours and the As interstitial (dots, above about 85 MHz which are not connected by solid lines) of the EL2 defect. It is the figure 2 of ref. [9], rotated bei 90°. Note, that the section of angles measured in a {110} plane is different from that in figure 9. In the analysis given in ref. [9] it was assumed, that one As neighbour had a slightly different shf tensor giving rise to the dashed lines next to the solid lines, describing the single nuclei 1 and 2. It was thought to have a slightly different isotropic shf constant. However, our new and better calculation of the effect of pseudo-dipolar coupling showed, that this assumption is not necessary: the line splitting can be explained by pseudo-dipolar coupling as in the case of the isolated As_{Ga}. Comparison of the ENDOR spectra in figures 7(a) and 7(b) already showed, that the nearest neighbours have the same lines for the same orientation in both defects. Both defects have also the same angular dependence. This is a surprising result. Thus, in table 3, the data of ref. [8,9] for the nearest neighbours should be replaced by those of the isolated As_{Ga}.

Within experimental error they are identical. However, for the EL2 defect in addition to the ODENDOR lines of the first As shell other lines were observed in the frequency range from 85 MHz to 130 MHz (see figure 7(b) and figure 13 for $B||[001]$). Their intensities are lower by a factor of 4-8 and their line widths are larger compared to those of the four nearest neighbours. From the measurement of the magnetic field dependence of these additional ODENDOR lines it was possible to identify them to be due to interactions with ^{75}As. The lower intensity of these ODENDOR lines seems to indicate that they belong to only one single nucleus.

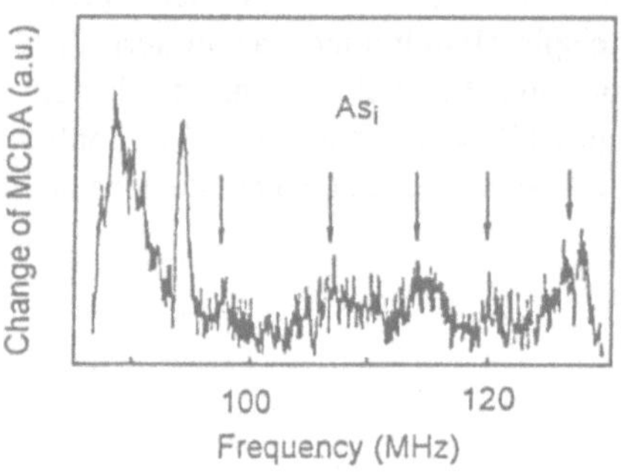

Figure 13
ODENDOR spectrum of the EL2 defect in the frequency range from 85 to 130 MHz for $B||[001]$ with additional As lines assigned to an interstitial As nucleus, marked by arrows.

These additional ODENDOR lines were detected in the whole MCDA spectrum of the EL2 and in all four ODEPR transitions. This shows that they belong to this defect and not to another defect, the spectrum of which could be superimposed to the EL2 ODEPR spectrum in lower intensity.

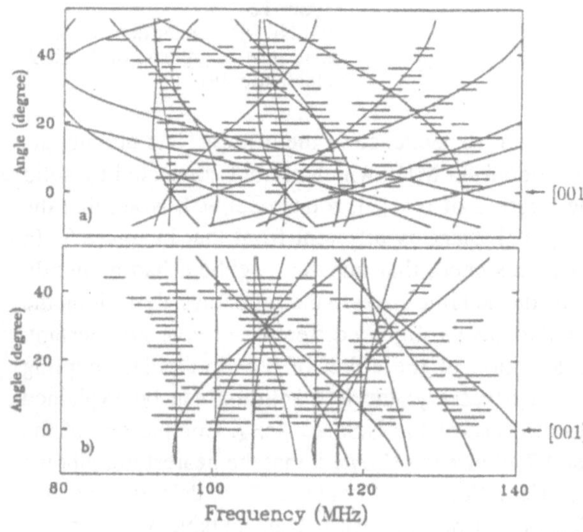

Figure 14
Angular dependence of the ODENDOR lines of the additional As interactions of the EL2 defect. The experimental line positions of the weak and broad lines are represented by horizontal bars; (a) with a theoretical calculation assuming that the axial interaction tensors are orientated parallel to <111> directions; (b) with a theoretical calculation assuming that the interaction tensors are oriented approximately parallel to <110> directions ($\theta = 30°$).

The ODENDOR angular dependence of these additional lines is shown in figure 14 for rotating the crystal in an {110} plane, beginning at $B||[001]$. Because of the rather low signal-to-noise ratio the line positions are represented by bars rather than squares which indicate the half widths of these lines. For simplification, the ODENDOR lines of the nearest neighbours were omitted. The analysis of the angular dependence is shown in figure 14 by the solid lines. Because of the large line width and the low signal-to-noise ratio it is difficult to arrive at an unambiguous and precise analysis of the angular dependence of these additional As lines. Figure 14(a) shows a calculated angular dependence assuming both shf and quadrupol tensors to be axial and to be oriented parallel to a <111> direction. This was the original analysis given in ref. [8,9]. Figure 14(b) shows a calculated angular dependence assuming the tensor orientations to be approximately along <110> directions (the calculation was done for $\theta \approx 30°$ off the <111> direction) with the interaction constants also given in table 3. It is not obvious, which of the two assumptions explains the lines better. Also for $\theta = 90°$ (from <111>) and $a/h = 195$ MHz, $b/h = 44$ MHz, $q/h = 4$ MHz a reasonable explanation of the data could be obtained. Thus, there is no doubt, that an additional ^{75}As shf interaction is there which does belong to the EL2 centre, but the precise determination of the shf and quadrupole tensors is not possible.

The anti-structure pair

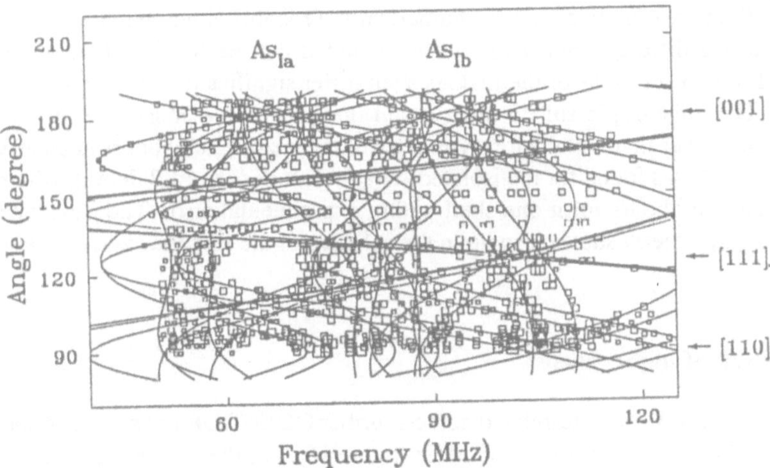

Figure 15 Experimental angular dependence of the first As shell of the anti-structure pair and a theoretical calculation (solid lines) indicating two different subshells As_{Ia} and As_{Ib} with <111> tensor symmetries. Pseudo-dipolar coupling is neglected.

The angular dependence of the ODENDOR lines of the nearest ^{75}As neighbours

Table 4 Superhyperfine and quadrupole interaction constants of the anti-structure pair, As_{Ga}-Ga_{As} (nnn).

	a/h [MHz]	b/h [MHz]	Θ_b	q/h [MHz]	Θ_q
1st As shell			<111> $\hat{=}0°$		<111>= 0°
nuclei 2-4	205.4 ± 0.2	50.8 ± 0.2	0° ± 2°	11.1 ± 0.2	0° ± 2°
nucleus 1	158.5 ± 0.2	54.7 ± 0.2	0° ± 2°	12.9 ± 0.2	0° ± 2°
Ga			<110> $\hat{=}0°$		<110> $\hat{=}0°$
^{69}Ga	22.5 ± 0.2	1.9 ± 0.2	19° ± 2°	0.9 ± 0.2	30° ± 2°
2nd As shell			<110> $\hat{=}0°$		<110> $\hat{=}0°$
	23.5 ± 0.2	2.4 ± 0.2	25° ± 2°	0.5 ± 0.2	30° ± 2°

of the anti-structure pair in the frequency range from 30 to 130 MHz is shown in figure 15. Two subshells with interaction tensors with <111> symmetry could be identified, i. e. there are two different types of As ligands with different shf tensors which have both their principal axes (that with the largest interactions) approximately in the <111> direction. Small deviations from axial symmetry could not be resolved. As for the ODENDOR lines of the nearest neighbours of the isolated As_{Ga} and the EL2 defect, the lines of the nearest neighbours of the anti-structure pair are split by pseudo-dipolar couplings. In contrast to the additional ODENDOR lines of the EL2 defect, the observed lines here have always nearly the same intensities. The shf and quadrupole parameters are listed in table 4. For the calculation of the angular dependence the ENDOR Hamiltonian was diagonalized numerically. Probably three neighbours are equivalent and one different, but it cannot be excluded that each subshell contains two ligands. The two subshells of the first As shell differ significantly in the isotropic shf constant. The anisotropic shf constant b and the quadrupole constant q do not differ much. The lower than tetrahedral symmetry of the nearest neighbours is clearly different from what was found for the isolated As_{Ga} defect and the EL2 defect, where the four nearest neighbours form one shell within experimental error. The deviation from tetrahedral symmetry shows also up in the splitting of the so-called zero phonon line [11].

1.4 ODENDOR of higher shells

For the three different As antisite-related defects further ODENDOR lines were observed in the frequency range from 10 to 35 MHz (figure 16). For the isolated As antisite defect ODENDOR lines were measured only between 10 and 25 MHz (figure 16(a)), for the EL2 defect they extend to 35 MHz (figure 16(b)). For both defects these lines belong to interactions with ^{75}As nuclei, as determined by field shift measurements. For the anti-structure pair, apart from ^{75}As ODENDOR lines, additional lines due to the two Ga isotopes, ^{69}Ga and ^{71}Ga, were observed (figure 16(c)).

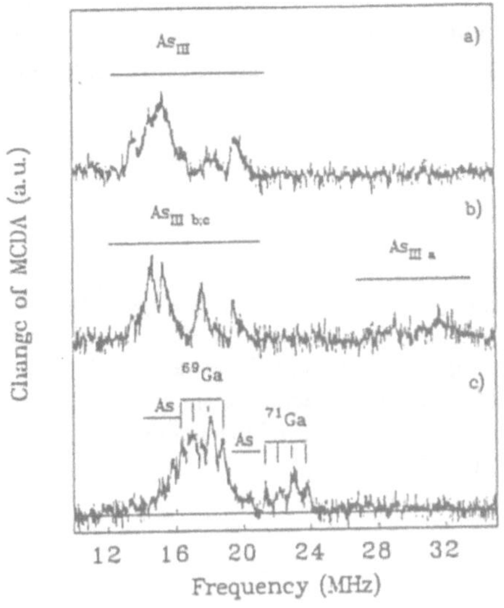

Figure 16
ODENDOR spectra of three different As antisite-related defects in the frequency range from 10 to 35 MHz measured at 1.5 K and $B||[001]$: (a) the isolated As_{Ga} defect, (b) the EL2 defect and (c) the next nearest anti-structure pair. The ODENDOR lines from ^{75}As belong to the second As shell. In the case of the anti-structure pair additional ODENDOR lines of the two Ga isotopes ^{69}Ga and ^{71}Ga were observed.

The isolated As antisite defect

The angular dependence of the further As lines of the isolated As_{Ga} defect is shown in figure 17. The ^{75}As lines between 12 and 22 MHz belong to the $m_S = -1/2$ state. The calculation of the ODENDOR angular dependence was done for an As shell which has <110> tensor symmetry (solid lines in figure 17).

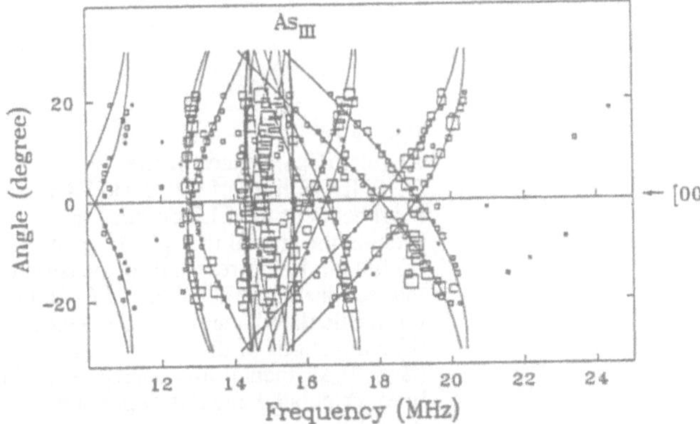

Figure 17
Experimental ODENDOR angular dependence of the second nearest As neighbours of the isolated As antisite defect (squares). The magnetic field was rotated in a {110} plane. The solid lines are the calculated angular dependences assuming <110> tensor symmetry (see table 3).

The nearest As with this symmetry would be in the third neighbour shell of the antisite (table 1). The lines are therefore assigned to this shell. The shf and quadrupole interaction constants are listed in table 3. The measured and calculated angular dependencies agree very well. The directions of the tensor angles in the crystal lattice are shown in figure 18 (see also table 3). The z-axes of the shf and quadrupole tensors point approximately into the direction of a neighbouring Ga nucleus (The angle would be 35.26° for this orientation), not into the direction of the antisite. This would have been expected in the simple approximation of classical point dipoles interacting, one nuclear dipole being located at the As nucleus of the second As shell, one electronic dipole at the As_{Ga} site.

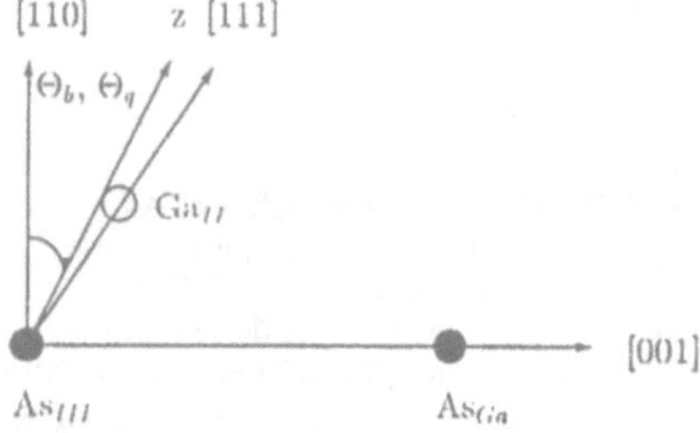

Figure 18
Orientation of the shf and quadrupole tensors Θ_b and Θ_q, respectively, for the second shell As neighbours (As_{III}). The measured tensor orientations of the second shell As neighbours point to the direction of the neighbouring Ga nuclei.

The EL2 defect

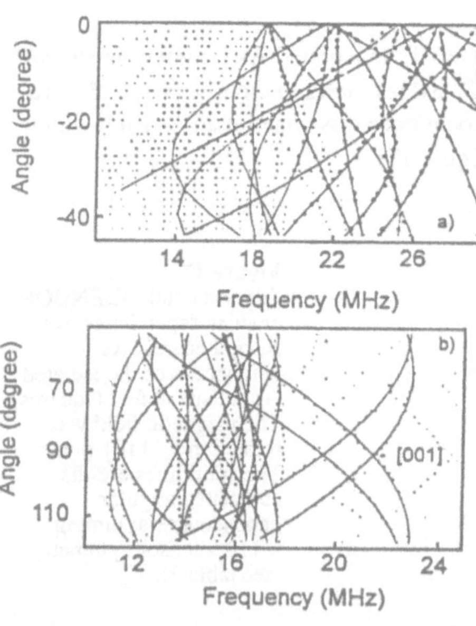

Figure 19
Angular dependences of the ODENDOR lines of the second As neighbours of the EL2 defect for rotating the crystal in a {110} plane. (a) full circles correspond to measured interactions of the subshell III_a (solid lines: calculated angular dependence); (b) interactions of the subshells $III_{b,c}$ (dots: experimental line positions; solid lines: calculated angular dependence). After [9].

In the frequency range between 10 and 35 MHz the EL2 defect has more lines than the isolated As_{Ga} defect (compare figures 16(a) and 16(b)). Both defects have lines between 10 and 22 MHz (labelled As_{III} for As_{Ga} and $AsIII, b, c$ for EL2 in figure 16), but EL2 has also lines with smaller intensity around 30 MHz, where nothing is found for the As_{Ga} defect. Figures 19(a) and 19(b) show the angular dependencies of these ODENDOR lines. The solid lines are calculated assuming <110> tensor symmetries. The shf and quadrupole interactions are listed in table 3. The angular dependence of the ODENDOR lines of the EL2 defect (labelled $As_{IIIb,c}$) between 10 and 22 MHz is very similar to the one observed for the isolated As antisite defect. The shf and quadrupole parameters and the tensor orientations of the ODENDOR lines labelled As_{IIIa} deviate clearly from the other ones. Thus, in the EL2 defect the second nearest neighbours do not form one shell as in the case for the isolated As_{Ga}, they form altogether three shells. One shell has clearly different tensors from the other two, which are almost identical and very similar to the one shell found for the isolated As_{Ga}. This shows that the EL2 defect has a symmetry which is lower than tetrahedral. No ODENDOR lines were observed which could be assigned to a Ga shell.

The anti-structure pair

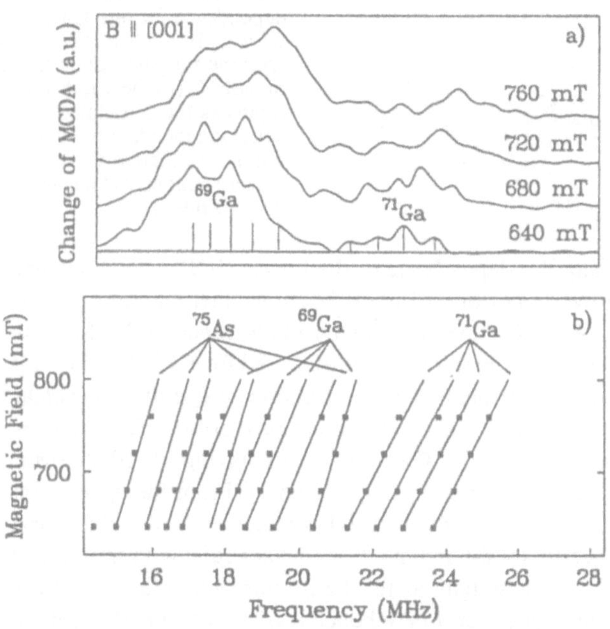

Figure 20
(a) ODENDOR spectra of the anti-structure pair in the frequency range between 14 and 28 MHz for $B||[001]$ measured for different magnetic fields. (b) Squares mark the frequency positions of the ODENDOR lines as a function of the magnetic field. The straight lines show the linear shift of the ODENDOR lines with the magnetic field, their slopes are proportional to the nuclear g_I factors of ^{75}As, ^{69}Ga and ^{71}Ga, respectively.

A new feature was observed for the anti-structure pair in the frequency range between 10 and 35 MHz. Apart from As interactions there are additional ODENDOR lines due to the two Ga isotopes, ^{69}Ga with 60 % abundance and ^{71}Ga with 40 % abundance (figure 16(c)). The assignment to Ga was again made with measurements of the magnetic field dependence of the ODENDOR lines (figure 20(a)) [11]. In contrast to the ODENDOR lines of the first neighbours a first order perturbation calculation (equation 10) sufficiently explains the observed line shifts. The result is shown in figure 20(b) as the straight lines, the squares represent the observed ODENDOR line positions. Both the intensity ratios of the ODENDOR lines and the magnetic field shift of their frequencies agree very well with their assignment to the two Ga isotopes. Only ODENDOR lines of the $m_s = -1/2$ state appear in this frequency range.

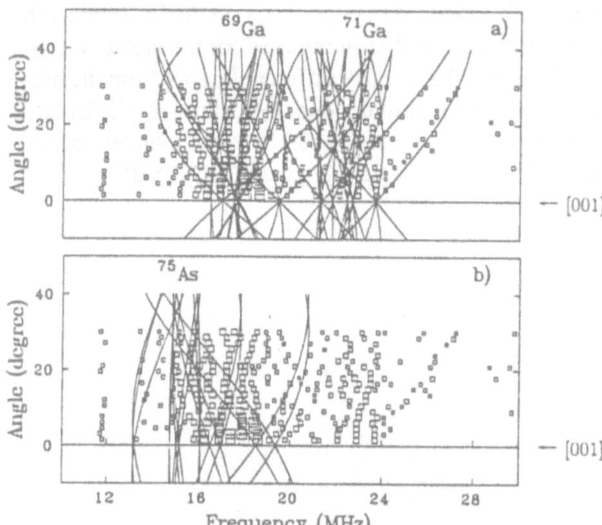

Figure 21
Angular dependence of the ODENDOR lines of the anti-structure pair in the frequency range from 10 to 28 MHz for a rotation of the crystal in {110} plane. The squares mark the frequency positions of the ODENDOR lines. (a) The solid lines are the calculated angular dependence for one Ga shell with <110> tensor symmetry with the shf parameters listed in table 4; (b) the solid lines represent the calculated angular dependence for the second As shell with <110> tensor symmetry (one shell was assumed).

The precise analysis of the angular dependence of the ODENDOR lines was very difficult because of the strong overlap of lines. Figure 21 shows their angular dependence. The lines could not be due to a Ga interstitial in a <111> direction. In this case only three lines (if there is a quadrupole interaction) are expected for $B||[001]$ since all nuclei would be equivalent in this direction. From the magnetic field shift it was clear that at least four Ga lines are resolved for each isotope (figure 20(a)). The analysis of the angular dependence agrees well with Ga interactions having <110> tensor symmetry as shown in figure 21(a) by the solid lines (table 4). From the agreement with the experiment follows that one type of Ga ligand is either on a site of the first Ga shell or on a second shell As site. Its shf and quadrupole axes (that with the largest interaction) point nearly to the next neighbour as for the As interaction tensors of the isolated As_{Ga} defect or the As interaction tensors (labelled $As_{IIIb,c}$) of the EL2 defect. There is only one Ga shf tensor for each isotope occurring in 12 <110> orientations. The interaction tensors were assumed to be axially symmetric. A more precise analysis detecting deviations from axial symmetry was not possible. The angular dependence of the As interactions in this frequency range (<110> tensor symmetry assumed) could be explained similarly as that for the isolated As_{Ga} defect (figure 21(b)). The shf and quadrupole interactions (table 4) are very similar to those found for the isolated As_{Ga} and to the As interactions (labelled $As_{IIIb,c}$) of the EL2 defect. We cannot say if these As interactions are split into different subshells with slightly different shf and quadrupole tensors as for the first As shell of the anti-structure pair because of the strong overlap of lines.

2 Interpretation of the ODENDOR results: structure models

2.1 The isolated As antisite defect

For the isolated As_{Ga} defect two different ODENDOR line groups were observed. Both belong to interactions with As nuclei. The ODENDOR line group in the frequency range from 35 to 120 MHz comes from one type of As interaction tensors with axial <111> tensor symmetry. The observed line splittings could be explained by pseudo-dipolar coupling between the nearest neighbours. The ODENDOR lines in the frequency range from 10 to 25 MHz are from As nuclei on {110} mirror planes ("<110>" symmetry). The observed interaction tensor symmetries are those expected for an isolated As antisite defect: <111> symmetry for the first As shell and <110> symmetry for the second As shell (see table 1). However, from the ENDOR spectra alone it was not possible to distinguish whether the As antisite atom is surrounded by four nearest neighbours or only by three neighbours as long as the geometry remains the same. To decide this one has to calculate the EPR line shape from the shf parameters. The EPR line width is sensitive to the number of interacting nuclei. For the calculation of the EPR spectrum all measured isotropic hf and shf interactions were taken into account in second order perturbation theory for a [001] crystal direction. The calculated half width of one hf EPR line is 32 mT, in excellent agreement with the measured one (see table 2) [10]. This means that all dominant ODENDOR interactions were measured and that both the ENDOR spectra and the EPR spectra are explained by the model of an isolated As_{Ga} defect. Note, that without taking into account the 12 third shell As-neighbours one calculates an EPR spectrum which shows a resolved shf structure of 13 lines in each hf line from the 4 nearest neighbours. As width of a single EPR transition that of the ENDOR lines ($\approx$ 2 MHz) was assumed for the EPR line shape calculation.

Whether or not this As antisite defect is "isolated" in the strictest sense, one cannot say, as long as a perturbation has a weak influence on the ODENDOR line positions of the first and second shell, i.e. a weak distortion in the first shell may be invisible because of the strong pseudo-dipolar coupling effects. Within experimental error this As antisite defect has the full tetrahedral symmetry. From its corresponding MCDA spectrum and its theoretical explanation it was concluded that the isolated As antisite defect has been identified in electron-irradiated SI GaAs kept below 77 K [10]. The microscopic structure model is shown in figure 22(a).

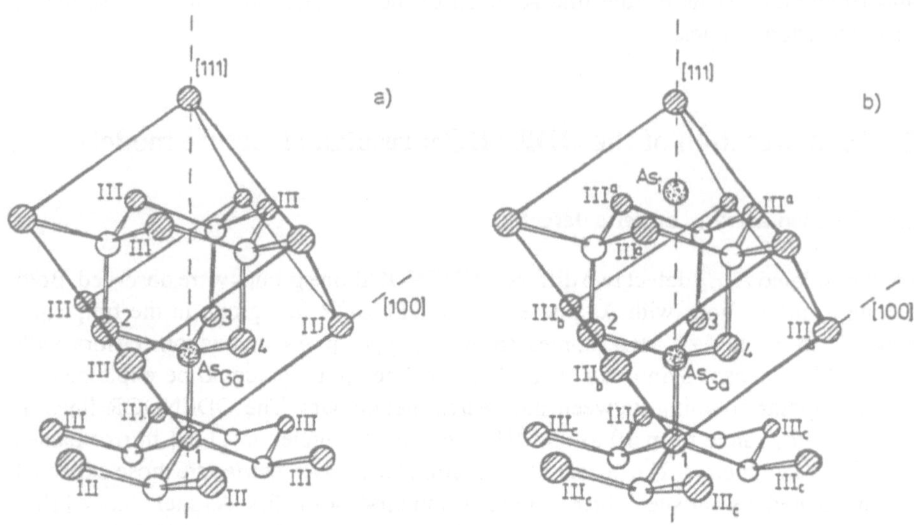

Figure 22 Structure models of two As antisite-related defects: (a) the isolated As_{Ga} defect, (b) the EL2 defect as an As_{Ga}-As_i pair defect (from ref. [9]).

2.2 The EL2 defect

For the EL2 defect four different ODENDOR line groups were measured, all belonging to interactions with As nuclei. The analysis of the ODENDOR angular dependencies yielded two different large As interactions, one with <111> tensor symmetries, the other possibly with <111> symmetry, possibly with lower symmetry and at least two As interactions with <110> tensor symmetries. The interpretation of the ODENDOR results that the EL2 defect is an As_{Ga}-As_i pair defect was based on the following arguments. The intense As ODENDOR lines in the frequency range from 35 to 120 MHz (figure 7(b), figure 12) were assigned to the four nearest As neighbours of the antisite [8,9]. They are within experimental error identical to those observed for the isolated As_{Ga} defect. The additional low intense As ODENDOR lines in the frequency range from 80 to 140 MHz (figures 13 and 14) were originally interpreted as being due to an additional As interstitial on a T_d position [8]. The influence of the interstitial onto the four nearest neighbours is apparently astonishingly small. The splittings of the ODENDOR lines of the first As shell are explainable by pseudo-dipolar coupling effects. This applies also to the "satellite" lines (dashed lines, see figure 12) which were interpreted previously as lines from one non-equivalent nearest neighbour. A slight distortion was assumed resulting from the presence of the interstitial As [8,9]. Our new measurements and calculations, however, show that it is not necessary to assume that one of the first neighbours has a slightly different interaction. For the four nearest neighbours of the antisite in the EL2 defect within experimental error no distortion is detectable. Therefore, a C_{3v} symmetry cannot be deduced from the interactions of the first shell as was thought before [8, 9]. If an As interstitial is present on a T_d site, it can be expected that the second As shell is split into different subshells [8,9]. The charge state of the interstitial must be such, that the resulting electron spin is zero, i.e. As^+, As^- or As^{3-}. Otherwise, both constituents of the pair would be paramagnetic and would have either a magnetic dipole-dipole interaction (estimated to be approximately 50 mT) or would result in a spin singlet or triplet state due to exchange interaction. This is not consistent with the observed (OD)EPR spectrum [9]. From an estimate of the quadrupole interaction of the As interstitial it was concluded that the interstitial is at a second nearest tetrahedral interstitial site ($d = 4.88$ Å). The influence of the interstitial on to the second nearest As shell splits it into three subshells (As_{IIIa}, $As_{IIIb,c}$), two could be analysed. The shf parameters and the quadrupole parameters of the shell $As_{IIIb,c}$—that is of those ligands which are far away from the As interstitial—are very similar to the corresponding As ligands of the isolated As_{Ga} defect. The resulting structure model of the EL2 as an As_{Ga}-As_i pair defect is shown in figure 22(b) [8, 9] where the As_i was placed into the <111> direction opposite to an As neighbour of the first shell.

In a recent paper it was mentioned that the As_i could also be in the "backbonding" direction from the As_{Ga} [22]. Then, the ODENDOR lines in the frequency range from 24 to 35 MHz could be assigned to the three nearest As of the third As shell (shell As_{Va} in [22]), which has also <110> symmetry (table 1). In principle, with ENDOR only the symmetry, principal values and orientations of the shf and quadrupole tensors can be determined. If the charged As_i was in the opposite direction it would need to be at a site far away, otherwise one would expect the quadrupole interaction constants of the four nearest neighbours to become too large. As was explained before the axial <111> symmetry of the As_i shf tensor assumed previously [8,9] is not a 'hard' experimental fact in view of the broad and weak ENDOR lines. The shf and quadrupole tensors could have another orientation and be non-axial.

Some difficulties arise from the simulation of the ODEPR line widths of the EL2 defect as an As_{Ga}-As_i pair with the shf parameters as measured by ODENDOR. A calculation of the line widths in second order, as was done for the isolated As_{Ga} defect, results in a half width of 37 mT for $B||[001]$. This is too large compared to the measured half width of 34 mT. A half width of 36 mT is obtained if the principal axis of shf tensor of the interstitial As is approximately parallel to a <110> direction (see table 3). As discussed above (see figure 14) such a tensor orientation cannot be excluded from the ENDOR angular dependence. Therefore, from the analysis of the ODENDOR spectra alone, without any theoretical interpretation or assumptions, one cannot determine the site of the interstitial As. Had there been better ENDOR signals, which would allow an unambiguous analysis of the angular dependence, one would probably be in a better position.

In order to demonstrate, that also a quite different defect model could be compatible with the EPR/ENDOR data, the following model is discussed. Assume, that As neighbour No. 1 of the first As shell is displaced in a <111> direction indicated as dashed

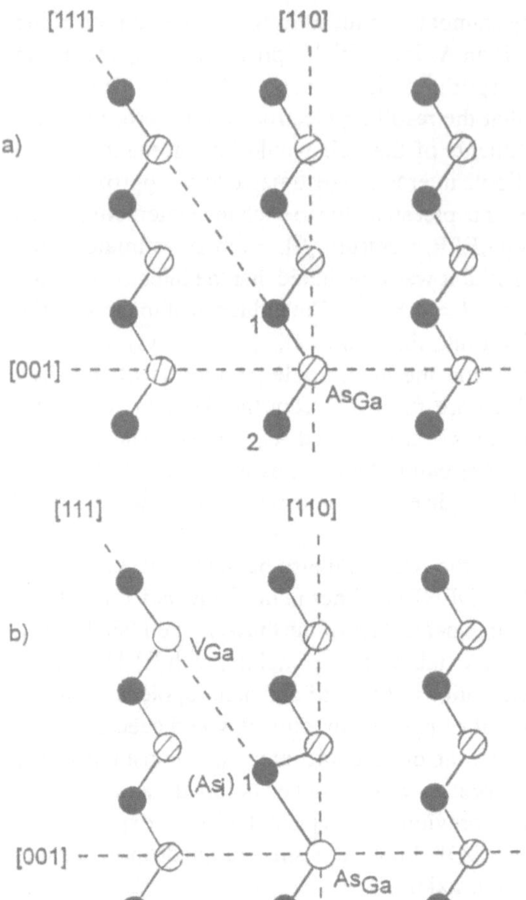

Figure 23
Structure model of the isolated As_{Ga} defect in a {110} plane. (a) a strechted As antisite defect where one As neighbour could be diplaced in a <111> direction. (b) Streched As model as a possible model of the EL2 defect. A Ga vacancy in the <111> direction could be responsible for the displaced As neighbour (see text for details).

line in figure 23(a). This displaced neighbour could be responsible for the additional ODENDOR lines as long as the other three As neighbours retain their <111> tensor symmetries and retain their shf and quadrupole parameters as for the As neighbours of the isolated As antisite defect. Of course, the hf constant of the central nucleus must also remain. In this case the displaced neighbour would be a quasi-interstitial. From ODENDOR the two possibilities are not distinguishable unless one is able to correctly calculate the pseudo-dipolar coupling with its angular dependence precisely for the cases with 3 and 4 equivalent neighbours and finds detectable differences in the ENDOR spectra for these two possibilities. So far this could not be done as mentioned above. The influence of the displaced neighbour on the second As shell could be the same as that of an additional As interstitial. The calculated EPR line widths with $a/h = 215$ MHz, $b/h = 44$ MHz from the values of "As_i" in <111> now yields 34 mT in agreement with the observed line width. The linewidth is now smaller since one large As shf interaction is missing compared to the As_{Ga}-As_i model. A possible explanation of such a strechted As antisite, where one As neighbour is displaced in the <111> direction, could be a diamagnetic Ga vacancy in this direction, probably in the V_{Ga}^{-} state (figure 23(b)). Also a divacancy may be involved in this defect similarly as the one suggested in ref. [23].

However, two arguments against the model of such a strechted As antisite are the following: the identical interactions of the first As shell of the isolated As_{Ga} and the EL2 defect and the non-observation of a splitting of the so-called ZPL of the EL2 defect inspite of the clearly lowered symmetry (trigonal) compared to the splitting of the ZPL of the anti-structure pair [11]. One would have expected the remaining three equivalent As neighbours to relax differently as those of the isolated As_{Ga} with the result of different shf and quadrupole interactions. This argument is supported by the observation, that different As interactions were observed for the anti-structure pair, where the "defect", the Ga, is out on the third shell. The possibility that the additional As_i low intense ODENDOR lines of the EL2 defect arise from another defect which may be hidden under the ODENDOR lines of the isolated As_{Ga} defect can be clearly ruled out. In totally different samples always the MCDA, the ODEPR and ODENDOR lines of the EL2 defect were observed. The low intense ODENDOR lines were measured in all three MCDA transitions (figure 6(b)) and in all four ODEPR hf-lines (figure 5(b)).

2.3 The anti-structure pair

For the anti-structure pair three different ODENDOR line groups were measured and analysed. Compared to the EL2 and to the isolated As antisite defect, the first As shell is clearly split into two different As interactions with <111> tensor symmetries. There is a clear deviation from tetrahedral symmetry here which also shows up in the splitting of the so-called ZPL [11].

The new feature in this defect was the observation of Ga ODENDOR lines. Neither in the EL2 [8] nor in the isolated As_{Ga} defect [10] any lines from a Ga shell have been observed. Thus it seems not very reasonable to assign the Ga lines to the first Ga shell. Furthermore, since the nearest As neighbours are clearly distorted from T_d symmetry, one would expect that the Ga shell is also broken into different subshells by the perturbation, which is the reason for the breaking of the T_d symmetry also of the first shell. However, this would lead to many more Ga ODENDOR lines than we have observed. A Ga vacancy in this shell, which may cause the occurrence of spin density in the Ga shell, would also lower the symmetry to several Ga subshells. Because only one Ga interaction with <110> tensor symmetry was measured, only one Ga atom can participate in this defect except for the Ga ligands in the Ga shell, which must have a vanishing or very small shf interaction. The nearest location of the extra Ga atom would be a next nearest neighbour (nnn) position in the second As shell. The resulting defect is the next nearest anti-structure pair As_{Ga}-Ga_{As} (nnn) which is shown in figure 24 [11].

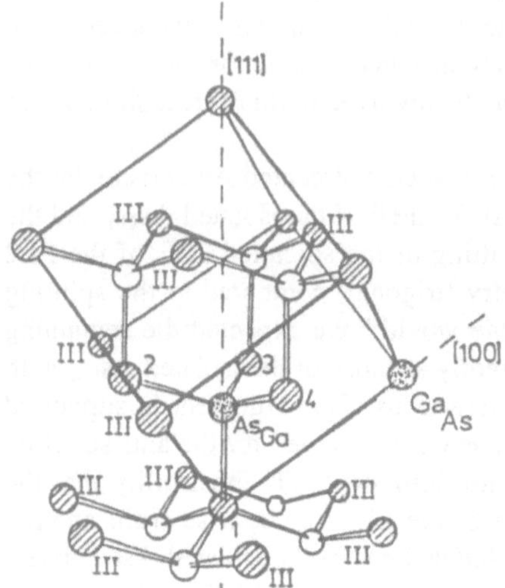

Figure 24
Structure model of the next nearest anti-structure pair, As_{Ga}-GaAs (nnn), where the involved Ga must be diamagnetic.

The angular dependence of the second As shell (<110> symmetry) could be explained similarly as that for the isolated As_{Ga} defect and the EL2 defect. The shf interactions (table 4) are very similar to those found for the isolated As antisite defect and to the shells IIIb,c of the EL2 defect, which are not influenced by the As interstitial. We cannot say whether the second As shell is split into subshells with slightly different shf tensors. If there is a difference, it is much smaller than that between shell III_a and shells $III_{b,c}$ of the EL2 defect [8].

3 Discussion

From the MCDA and ODENDOR measurements it is clear that the three defects must have different microscopic structures, which all contain the As_{Ga} as a constituent with approximately the same spin density at the As_{Ga} nucleus (i.e. same hf interaction). The spectra assigned to the isolated As_{Ga} reflect the highest symmetry, which within experimentel error is T_d. The surprising result is, that the ENDOR spectra of the nearest As neighbours of $EL2^+$ are not distinguishable from those of the isolated As_{Ga}, although the second shell As has a lower symmetry and there is one different large ^{75}As shf and quadrupole interaction not found in the isolated As_{Ga}. The lower symmetry is also reflected in the MCDA spectra: one finds a single derivative structure for the isolated As_{Ga} as expected for a simple $A_1 - T_2$ transition, while the more complicated pattern of EL2 reflects a splitting of the excited states [10]. Many other structure sensitive experiments were done in order to get a microscopic model of the EL2 defect. The piezo-splitting of the so-called zero phonon line (ZPL) observed in the optical absorption was interpreted for a defect which has the full tetrahedral symmetry [2]. The underlaying assumption was that the ZPL transition is an internal $A_1 - T_2$ transition. This interpretation must be challenged because of the experiments done by Baj and Dreszer [24], in which the ZPL and the 1.18 eV absorption bands shift in the opposite sense under hydrostatic pressure. Another interpretation is that the sharp line is connected with a transition to the L point of the conduction band. The observed splitting is the result of the symmetry of this conduction band minimum [25-27]. Therefore, from these experiments no arguments about the symmetry of the ground state configuration of the EL2 defect can be safely derived.

Investigations of sharp transitions in photoluminescence under uniaxial stress and in magnetic fields, which were assumed to be connected with the EL2 defect, however, favour the model of the isolated As antisite defect [28]. These experiments seem to contradict our ENDOR results. We have found that the isolated As antisite defect can only be produced in low temperature electron-irradiated SI GaAs which contains EL2 defects before irradiation and that the EL2 defect has lower than T_d symmetry. The better resolution in the ENDOR experiments (50 neV) compared to the spectral resolution in the photoluminescence investigations (500 neV) could be an explanation for the apparent contradiction. On the other hand, if the photoluminescence probes only the positions of the nearest neighbours in the ground state, then from ODENDOR an apparent T_d symmetry is understandable, since also in ENDOR the nearest neighbours appear to be tetrahedrally located around the As_{Ga}.

The theoretical explanation of the transition to the metastable state as a transfer of the As antisite atom into an interstitial position following a bond rupture was calculated for the model of the isolated As antisite defect [4,5]. The question arises, however, whether a complex defect which involves an As antisite can also have such a transfer. All three As antisite-related defects, which were investigated in this study, show metastable properties. The optical processes and the transition probabilities differ between the different As antisite species, however, the thermal recovery temperature is with about 140 K always nearly the same. We think that these observations are an indication for the same fundamental process. Details of the optical properties and the transfer into the metatstable states of the three different As antisite-related defects will be published elsewhere [29].

Many other structure sensitive experiments led to the conclusion that the EL2 defect is an As antisite-related complex, some determine the symmetry to be C_{3v} [30-33]. This symmetry was also concluded previously from ODENDOR spectra [8,9], which were interpreted to originate from an As_{Ga}-As_i pair defect where the As_i is nearly two bond lengths away from the antisite. For this model for EL2 a low binding energy between As_{Ga} and the As_i was calculated which seems to be in disagreement with the high thermal stability. However, nobody knows the thermal stability exactly. Rapid quenching from temperatures higher than 950 °C destroys EL2 defects (ITC treatment) [34]. After such a destruction one cannot measure the isolated As_{Ga}. Thus not only the complex partner of the As_{Ga} is removed. EL2 can, however, be reformed by annealing of the sample in a temperature range from about 650 °C to 800 °C. If there is a very mobile component, the As_i for example, then the EL2 defect as an As_{Ga}-As_i pair can apparantly be rebuilt at lower temperatures than that of the EL2 "stability temperature" of 950 °C. It should be noted that the EL2 concentration was not yet measured at elevated temperatures. Thus, at say 700 °C or 800 °C there may be a small concentration present as a result of a dynamical equilibrium between destruction and reformation, while at 950 °C this concentration is Zero. Note, that slow cooling from 950 °C does not destroy EL2! One obtains the usual concentration at low temperature. Only rapid quenching, which does not allow the reformation to happen, destroys EL2. Never the isolated As_{Ga} was found after any sort of heat treatment in as-grown SI GaAs, inspite of extensive attempts to detect it in our laboratory.

In two recent theoretical publications [35,36] different As antisite-As interstitial configurations were discussed as possible candidates for the EL2 defect. Chadi [35] has found in his calculations a low energy structure of the interstitial in which it is bonded in a twofold co-ordinated bridge and negatively charged. Together with a positively charged antisite this pair gets a low binding energy from 100 to 200 meV because of Coulomb attraction. In this model the interstitial position deviates from the <111> axis. In principle, from the ODENDOR measurements as discussed above, such a model cannot be excluded because of the rather low signal-to-noise ratio of the additional As lines. However, there are some difficulties in the understanding of such a pair which were mentioned by others [36]. The neutral charge state of this pair $(As^{+}_{Ga}$-$As^{-}_{i})^{0}$ would be responsible for the midgap level and EPR active, the metastable state would be the As^{0}_{Ga}-As^{-}_{i} state. From our point of view these charge states seem not to be consistent with what is observed experimentally. Zhang and Bernholc [36] have also calculated different As_{Ga}- interstitial complexes. They found no remarkable binding energies for such complexes. Thus, they concluded that the isolated As antisite defect would be the best candidate for the EL2 defect. They tried to explain the ODENDOR results for two different As antisite-related defects, the dominant one would be the isolated antisite. From the MCDA and ODENDOR measurements this possibility, as mentioned before, seems to be very unlikely. The ODENDOR spectra measured for EL2 could be observed in the whole MCDA (figure 6b) and in all four ODEPR transitions (figure 5(b)) of the EL2 defect. The typical MCDA and ODENDOR spectra were found in totally different GaAs materials including after electron irradiation at low temperature and annealing to 520 K! Therefore, we conclude that the EL2 defect as measured by the optically detected magnetic resonance is an unique defect and has an overall symmetry lower than T_d.

Now, there arises the question whether the proposed structure model of the EL2 defect as an As_{Ga}-As_i pair [8,9] is the only possibility to interpret the ODENDOR measurements. From thermodynamical considerations of the origin and destruction of the EL2 defect in melt-grown and epitaxially grown GaAs Morrow [37] proposed an atomic model for the EL2 defect having the structure of an As_{Ga}-V_{Ga} pair. Also an As_{Ga} divacancy pair was considered as a possible candidate for the EL2 defect [23]. From the ODENDOR measurements, as mentioned before, the additional As lines in the frequency range from 80 to 140 MHz can be interpreted as resulting from one special As neighbour of the first As shell as long as the other three As neighbours retain their positions and shf interaction constants identical within experimental error with those of the isolated As_{Ga} defect. These three neighbours in four centre orientations would yield the same angular dependence as for the whole shell except possibly for discernable effects of pseudo-dipolar coupling, which could not yet be calculated exactly as mentioned above. From the estimate of the quadrupole constant of the special neighbour one would expect that this neighbour is displaced in a <111> direction far from the antisite. Responsible for such a displacement could be a Ga vacancy in this direction, i.e. a Ga vacancy from the sixth shell, 9.76 Å far away from the antisite. The resulting pair would be a strechted As antisite complexed with a Ga vacancy, $As_{Ga}As_3As_1V_{Ga}$. This pair would have C_{3v} symmetry. Also the ODEPR line width could be explained within this model.

The remaining questions are: Does the model of EL2 as an $As_{Ga}As_3As_1V_{Ga}$ pair hold for the special properties of the EL2 like the metastability, the energy levels in the band gap etc ...? For answering these questions theoretical calculations of the atomic structure of such a model would be helpful. This structure model is as compatible with the ODENDOR data as is the As_{Ga}-As_i model. Only the investigations of the sharp transitions in photoluminescence under uniaxial stress and magnetic fields seem to contradict the strechted As_{Ga}-V_{Ga} model, since there is no T_d symmetry any more. From ENDOR alone one cannot decide which of the proposed models is realised. Clear is that the EL2 defect has an overall symmetry lower than T_d, but possibly with "tetrahedral" nearest neighbours.

It is surprising that the isolated As antisite is only observable after electron irradiation at low temperatures in crystals which contain EL2 defects before irradiation and that this defect is not stable near room temperature. It is very likely that the isolated As_{Ga} defect is the result of the destruction of EL2 defects caused by the irradiation.

In GaAs crystals, which were electron-irradiated at low temperature, another defect is prominent in MCDA and ODEPR independent of the conducting type. On the basis of EPR experiments this defect was recently attributed to the Ga vacancy in the (2-) charge state, trigonally distorted because of Jahn-Teller relaxations [38]. This defect is not very stable at room temperature, similarly as the isolated As_{Ga} defect. Upon warming the low temperature electron-irradiated GaAs samples to room temperature the anti-structure pair appears. We proposed that near room temperature an As atom from the second nearest As shell jumps into the Ga vacancy forming an As_{Ga}-V_{As} (nnn) pair defect. Interstitial Ga which is assumed to be mobile near room temperature [39], goes into the second shell As vacancy and the resulting defect is our anti-structure pair consisting of an As antisite and a Ga antisite separated by 4.88 Å [11].

Surprinsingly, after heating a SI GaAs sample, which contains EL2 defects prior to the irradiation, to about 520 K, the MCDA of the anti-structure pair disappeared and that of EL2 appeared. By ODENDOR it was established that indeed EL2 centres were measured. EL2 appears now in both its charge states ($EL2^0$ and $EL2^+$) with concentrations of $1.0 \times 10^{16}\,cm^{-3}$ and $5.1 \times 10^{16}\,cm^{-3}$, respectively (before irradiation it were 2.1×10^{16} cm^{-3} and 0.5×10^{16} cm^{-3}, respectively). Thus the total concentration of EL2 is about a factor of 2.5 higher than before irradiation. From this and the failure to see EL2 before we have concluded that EL2 was destroyed upon electron irradiation at low temperature and was formed at 520 K as a new pair defect. In a recent publication another group found EL2 concentrations at least one order of magnitude higher than before using a high dose electron irradiation at low temperatures and thermal annealing to 520 K [40]. In GaAs samples electron-irradiated at low temperatures, which do not contain EL2 defects before irradiation, i.e. in a Ga-rich sample and in a highly n-type Te-doped sample, neither the isolated As antisite defect and after heating to about 520 K nor EL2 defects could be observed in neither charge state. This may be an indication that the EL2 defect is the proposed As_{Ga}-As_i pair defect, because the As interstitials are assumed to be mobile near 520 K.

4 Conclusion

The EL2 defect is not an isolated As_{Ga} defect, but has lower symmetry, which is clearly seen in the MCDA spectrum as well as in higher neighbour shells in ENDOR. A surprising result is that both the isolated As_{Ga} as well as the EL2 defect are very similar if not identical with respect to their nearest neighbour interactions. The As_{Ga} ENDOR results show two shells of ^{75}As neighbours, one with large and one with small interactions, while the EL2 defect has two shells of ^{75}As neighbours with large interaction and three shells of ^{75}As neighbours with small interactions. The EPR line width is explained if there are either three "equivalent" neighbours in one shell and one extra nucleus with large interactions (stretched As_{Ga} model) or four equivalent nuclei as nearest neighbours and one extra nucleus as an interstitial. Because of the weak ENDOR lines of that extra ^{75}As it is not possible to unambiguously determine the orientation and principal values of its interaction tensors. The EPR line width favours slightly the "stretched As_{Ga}" model. The As_{Ga}-As_i pair model is a likely defect model for EL2, but one could imagine also other models compatible with the MCDA/ODENDOR data as discussed above. Here is the limit of ENDOR, since it is not a true microscope. ENDOR can only yield a set of superhyperfine and quadrupole tensors with their principal values and orientations, not more. The model finding requires interpretation. In the anti-structure pair the symmetry breaking of the nearest neighbours is obvious from the ENDOR spectra and the occurrence of a small Ga interaction is also beyond doubt. The proposed model seems a resonable interpretation.

The conventionally detected EPR spectra recorded always after room temperature electron irradiation of any type of GaAs (p type, n type, SI) [41-44] an As_{Ga}-related defect with large ^{75}As hf interaction. They must have been due to the anti-structure pair As_{Ga}-Ga_{As} (nnn) described above. This defect was originally also-called As_{Ga}-X_1 [11].

The $^{75}As_{Ga}$ hf interaction measured for these three defects is the highest found so far.

There are other As antisite related defects which have a reduced $^{75}As_{Ga}$ hf interaction. They can be produced by electron irradiation of n-type GaAs [45] or they are found in GaAs layers MBE-grown at 325 °C and 400 °C, in MOCVD grown $Al_xGa_{1-x}As$ or in n-type undoped Horizontal-Bridgeman-grown GaAs [45]. Their hf interaction constants are different and vary approximately between 1800 MHz and 2300 MHz, the line widths are larger than those discussed above ($\approx$ 50 mT). Clearly, these defects must be As_{Ga}-complexes where the unpaired spin density is more delocalised into the lattice, probably due to a much more pronounced "perturbation" compared to the cases discussed here. Unfortunately, not more can be said about their structure at present. It was speculated that an As vacancy is the nearest neighbour in the n-type electron-irradiated GaAs [41].

Upon plastic deformation As_{Ga} related defects are created, which have again the large hf interaction as has EL2 and the isolated As_{Ga} defect, but which differ in the spectral shape of the MCDA [46]. ODENDOR showed also that the atomistic structure of EL2 changes in deformed GaAs and that no new EL2 defects were created by the plastic deformation. However, the ENDOR lines were too broad to allow a more precise analysis of the spectra [46].

Bibliography

[1] J.-M. Spaeth, K. Krambrock and D. M. Hofmann, in *The Physics of Semiconductors* edited by E. M. Anastassakis and J. D. Joannopoulos (World Scientific, Singapore), Vol.1, p. 441 (1990).

[2] M. Kaminska, M. Skowronski and W. Kuszko, Phys. Rev. Lett. **55**, p. 2204 (1985).

[3] G. A. Baraff, Phys. Rev. B **41**, p. 9850 (1990).

[4] J. Dabrowski and M. Scheffler, Phys. Rev. Lett. **60**, p. 2183 (1988).

[5] D. J. Chadi and K. J. Chang, Phys. Rev. Lett. **60**, p. 2187 (1988).

[6] M. J. Caldas, J. Dabrowski, A. Fazzio and M. Scheffler, in *The Physics of Semiconductors* edited by E. M. Anastassakis and J. D. Joannopoulos (World Scientific, Singapore), Vol. 1, p. 469 (1990).

[7] H. J. von Bardeleben, D. Stievenard, D. Deresmes, A. Huber and J. C. Bourgoin, Phys. Rev. B **34**, p. 7192 (1986).

[8] B. K. Meyer, D. M. Hofmann, J. R. Niklas and J.-M. Spaeth, Phys. Rev. B **36**, p. 1332 (1987).

[9] B. K. Meyer, Rev. Phys. Appl. **23**, p. 809 (1988).

[10] K. Krambrock, J.-M. Spaeth, C. Delerue, G. Allan and M. Lannoo, Phys. Rev. B **45**, p. 1481 (1992).

[11] K. Krambrock and J.-M. Spaeth, Phys. Rev. B **47**, p. 3187 (1993).

[12] J.-M. Spaeth, J.R. Niklas and R.H. Bartram, in *Structural Analysis of Point Defects in Solids*, Springer Series in Solid-State Sciences, Vol. 43 (1992).

[13] A. Abragam and B. Bleaney, *"Electron Paramagnetic Resonance of Transition Ions*, Oxford, N. Y. (1970).

[14] T. E. Feuchtwang, Phys. Rev. **126**, p. 1628 (1962).

[15] H. Seidel, Habilitationsschrift, University of Stuttgart (1966, unpublished).

[16] J. Hage, J. R. Niklas and J.-M. Spaeth, Mater. Sci. Forum **10-12**, p. 259 (1986).

[17] D. E. Holmes, R. T. Chen, K. R. Elliot and C. G. Kirkpatrick, Appl. Phys. Lett. **40**, p. 46 (1982).

[18] N. Tsukada, T. Kikuta and K. Ishida, Jap. J. Appl. Phys. **24**, L689 (1985).

[19] D. Y. Jeon, H. P. Gislason, J. F. Donegan and G. D. Watkins, Phys. Rev. B **36**, p. 1324 (1987).

[20] J.-M. Spaeth and J. J. Lappe, Appl. Magn. Res. **2**, p. 311 (1991).

[21] H. Söthe and J.-M. Spaeth, J. Phys.: Condens. Matter **4**, p. 9901 (1992).

[22] K. Krambrock and J.-M. Spaeth, Mat. Sci. Forum **83-87**, p. 887 (1992).

[23] J. F. Wager and J. A. van Vechten, Phys. Rev. B **35**, p. 2330 (1987).

[24] M. Baj and P. Dreszer, Phys. Rev. B **39**, p. 10470 (1989).

[25] M. Skowronski, Mat. Sci. For., Vol. **104**, p. 405 (1987).

[26] H. J. von Bardeleben, Phys. Rev. B **40**, p. 12546 (1989).

[27] M. Lannoo, Semicon. Sci. Technol. **6**, B 16 (1991).

[28] M. K. Nissen, A. Villemaire and M. L. W. Thewalt, Phys. Rev. Lett. **67**, p. 112 (1991).

[29] K. Krambrock, M. Heße and J.-M. Spaeth, to be published.

[30] M. Levinson and J. A. Kafalas, Phys. Rev. B **35**, p. 9383 (1987).

[31] L. Dobaczewski, Mat. Sci. For. 38-41, Vol. 1, p. 113 (1988).

[32] J. C. Culbertson, U. Strom and S. A. Wolf, Phys. Rev. B **36**, p. 2962 (1987).

[33] T. Haga, M. Suezawa and K. Sumino, Japn. J. Appl. Phys. **27**, p. 1929 (1988).

[34] J. Lagowski, H. C. Gatos, C. H. Kang, M. Skowronski, K. Y. Ko and D. G. Lin, Appl. Phys. Lett. **49**, p. 892 (1986).

[35] D. J. Chadi, private communication (to be published in Phys. Rev B)

[36] Q.-M. Zhang and J. Bernholc, private communication (to be published in Phys. Rev. B)

[37] R. A. Morrow, J. Appl. Phys. **70**, p. 6782 (1991).

[38] Y. G. Jia, H. J. von Bardeleben, D. Stievenard and C. Delerue, Phys. Rev. B **45**, p. 1685 (1992).

[39] C. Corbel, F. Pierre, K. Saarinen, P. Hautojärvi and P. Moser, Phys. Rev. B **45**, p. 3386 (1992). [40] A. Pillukat and P. Ehrhart, Appl. Phys. Lett. **60**, p. 2794 (1992).

[40] N.K. Goswami, R.C. Newman and J.E. Whitehouse, Solid State Comm. **40**, p. 473 (1981)

[41] T.A. Kennedy, B.J. Faraday and N.D. Wilsey, Bull. Am. Phys. Soc. **26**, p. 255 (1981)

[42] R.B. Beall, R.C. Newman, J.E. Witehouse and J. Woodhead, J. Phys. C: Solid State Phys. **18**, 3273 (1985)

[43] H. J. v. Bardeleben, A. Miret, H. Lim and J.C. Bourgoin, J. Phys. C, Solid State Physics **20**, p. 1353 (1987)

[44] K. Krambrock, M. Linde, J.-M. Spaeth, D.C. Look, D. Bliss and W. Walukiewicz, Semicond. Sciene and Technology **7**, p. 1037 (1992)

[45] D.M. Hofmann, B.K. Meyer, J.-M. Spaeth, M. Wattenbach, J. Krüger, C. Kieselowski-Kemmerich and H. Alexander, J. Appl. Phys. **68**, p. 3381 (1990)

The Use of Effective Medium Theories in Optical Spectroscopy

Wolfgang Theiß

I.Physik. Inst. RWTH Aachen, Postfach o. Nr., 52056 Aachen

1 Summary

For the optical analysis of heterogeneous materials the microtopology of the samples plays an important role. In the long wavelength limit (i.e. light wavelengths much larger than the typical size of the inhomogeneities) effective medium theories give the desired connection between the component properties and the average 'effective' optical behaviour. On the basis of the general Bergman representation for effective dielectric functions we discuss simple and advanced effective medium concepts and show how they can successfully be used in optical spectroscopy.

2 Introduction

Most of the knowledge about solid state physics has been obtained preparing and studying very pure materials. In order to find rules and develop theoretical models the systems studied had to be as simple as possible, in theory as well as in experiments. Therefor an important basic step for many conceptual breakthroughs has been the preparation of large single-crystalline samples—minimizing discrepancies between theoretical simplifications and experimental conditions.

By now the understanding of many solid state phenomena observed in pure single-crystalline systems is quite complete and is the basis of numerous technical applications, the most prominent and important one being semiconductor electronics. In many cases, on the other hand, what is needed for an application is not exactly what is found in clean samples—then a change of the 'pure material properties' is demanded. Consequently semiconductor physics has learned to handle 'dirty' materials with the desired deviations obtained by doping pure materials on an atomic scale with impurities.

Besides this 'shifting' of a bulk property by disturbing a pure material one can also get new materials for technical applications by mixing different constituents above the atomic scale: in this case each component remains unchanged microscopically, but the composite material may show new combined features.

A good example is the use of porous materials for heat insulation purposes. When mechanical stability and low thermal conductivity is needed simultaneously one can try to combine a solid component which contributes stability and a gas which prohibits heat transfer. As one can easily imagine the overall performance of a solid-gas composite depends strongly on the way one puts the pieces together. Fig.1 shows two extreme cases of our new material filling the space between two walls: to the left a version optimized for mechanical stability is shown (with still quite high thermal conductivity) whereas to the right a good heat insulator with negligible stability is sketched. The dependence of the effective thermal conductivity on geometry can be used to vary it (in limited ranges) continuously and to achieve any desired value.

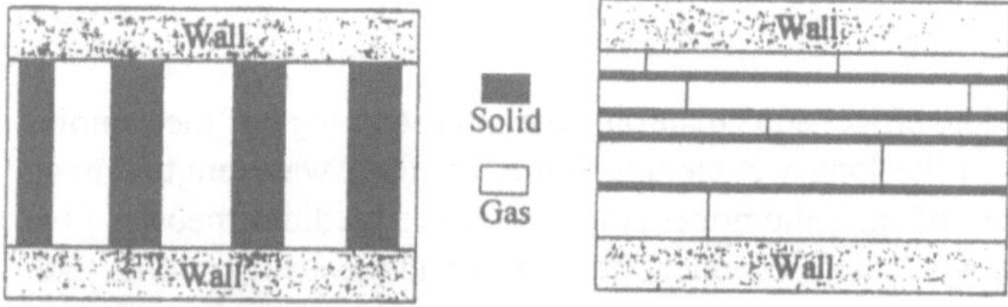

Figure 1 Heat insulation and mechanical stability between two walls obtained by a solid-gas composite material. To the left a topology prefering stability is shown, whereas the right geometry corresponds to a good heat insulator but is mechanically unstable.

Composite materials have gained interest recently not only for their technical applications but also due to a still unsolved problem in solid state physics, namely to find an adequate level of description for small particles. Concepts to treat atoms and molecules on one hand and large crystals on the other hand are well established, but an appropriate theory for crystallites with e.g. 10^6 or 10^8 atoms which would connect discrete molecular energy levels to continuous bands of large crystals does not exist yet. Experimental work in this regime is—in most cases—not possible considering only one particle but one must sum up contributions from a great number of them. Therefor a system of small clusters embedded in a host material is the typical case. Here—from an experimental point of view—one is tempted to collect signals from as many clusters as possible, i.e. to study rather dense systems, but this on the other hand complicates any evaluation of information due to the mutual cluster interaction which must be taken into account in dense systems.

Among the various properties of mixed materials their optical behaviour plays a special role—being interesting enough for itself it must be known for any nondestructive optical analysis. Since in general heterogeneous materials are more or less efficient light scatterers the determination of their optical properties is a very complex problem [1]. Nevertheless there are certain cases where quite successful approximative treatments have been developed. In very dilute systems incident radiation is absorbed or scattered by the individual particles—no interaction of the particles must be taken into account and the outgoing scattered light can be obtained as a sum of single-scattering contributions. In case of spherical particles everything can be explained perfectly by the Mie theory [2, 3].

For a treatment of dense systems more restrictions have to be applied: for wavelengths small compared to the mean inhomogeneity size one can neglect eventually the phase relation between partial waves scattered from different scatterers. Then the random-walk of light through such a system can be described by radiation transfer theories which do the necessary multiple scattering statistics [1, 4, 5].

The regime where light wavelength and inhomogeneities are of the same order of size is certainly the hardest for theoretical treatments—on the other hand it promises much information about microtopology since phase information about the scattering processes is still present in the outgoing radiation. The observation of enhanced backscattering and its explanation as a 'weak localization' effect [6] has encouraged a lot of work dealing with systems of this type. As to our knowledge still there is no established handy concept ready for applications.

This article is about the quasistatic limit where the light wavelength is much larger than the system inhomogeneities. In this case the light does not resolve the microscopic structure of the topology and 'sees' a quasi-homogeneous, a so-called effective medium (see fig.2). A light wave passing a characteristic section of such a system can be considered as—from a microscopic point of view—slowly increasing and decreasing electric fields. The polarizable particles can interact with each other by their induced electric fields on a time scale much smaller than that of the phase change of the radiation and the response can follow the external field instantaneously. Therefor a description on the basis of electrostatics is allowed.

The averaged optical properties of effective media in the long-wavelength limit are very much the same as those of homogeneous ones: a plane-parallel slab will show a reflected and a transmitted wave when illuminated by a single beam, scattered light can be neglected. Spectroscopy can be done with the usual methods developed for homogeneous samples. The only difference is that the dielectric response of a heterogeneous system must be described by a somehow averaged dielectric function— the so-called effective dielectric function. The kind of averaging to be employed depends on the microgeometry very much the same as the effective thermal conductivity in the introductory example depends strongly on topology (see fig.1).

In this article we restrict ourselves to two-phase composites since for many-component systems no satisfying theory exists. We distinguish formally between particles with dielectric function ε and matrix material with dielectric function ε_M although mixtures of two phase In particular if the particles build up a continuous network it is not clear which component is the matrix and which is embedded. The effective dielectric function is denoted in this article as ε_{eff}.

In a first step we will show by an example that simple 'averaging theories' widely used have their limitations and that an improvement is required.

In order to find the correct effective medium theory one has to take into account very

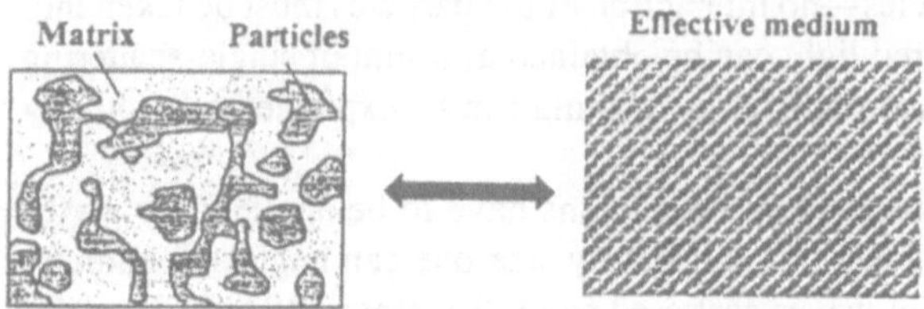

Figure 2
Principle of the effective medium concept: a heterogeneous two-phase composite (here represented as particles embedded in a matrix material) is replaced by a homogeneous 'effective medium' with averaged material properties.

flexible the system topology. This is demonstrated by a model system of interacting dipoles whose dielectric response depends strongly on their arrangement.

The necessary statistical treatment of the geometry influence for extended systems is possible using the so-called Bergman representation of effective dielectric functions. It plays the central role in this article and is discussed in detail. On the basis of the Bergman representation one can develop new simple effective medium concepts tailored for certain types of microgeometry.

Finally we turn to the problem of determining the dielectric function of small particles from measured effective dielectric functions. Here information about size-dependent changes of the bulk dielectric function is wanted but sometimes this effect cannot be separated from topology effects. This ambiguity can be removed partially by looking at the so-called rigorous bounds which enclose all effective dielectric constants being compatible with pure topology effects on the basis of bulk values for the particle material.

3 Simple effective medium concepts

As has been shown in the introduction the key problem in finding the effective dielectric function for a given system is to take into account the microtopology correctly. Many effective-medium approaches that have been developed in the past were obtained by very crude assumptions on the geometrical arrangement of the two mixed phases. The most simple ones are mentioned shortly in this section, where we consider a theory as simple when the description of the topology influence on the effective dielectric function is done using one or two parameters only.

The most prominent cases (due to Maxwell Garnett and Bruggeman, see below) use the volume fraction of the particle material as the only quantity characterizing the topology. One then can achieve very simple expressions for the effective dielectric function of the heterogeneous system which depends on the dielectric function of the constituents and the volume fraction. Moreover an inversion is usually straightforward permitting the determination of the dielectric function of the particles from measured effective optical properties. Because of their tempting simplicity these simple concepts are still being used frequently in optical spectroscopy—even in cases where the microgeometry under consideration does not meet the specifications needed for the applied formula. It will be discussed below in which cases the use of a simple formula is uncritical and in which cases wrong results should be expected.

To save space for improved effective-medium theories we do not discuss the derivations leading to the simple formulas given. They can be found in the cited literature.

3.1 Maxwell Garnett

A recipe for mixing dielectric functions of two components was given very early by Maxwell Garnett [7]. He found for systems of spherical particles being considerably far away from each other (this implies low volume fractions) that the relation

$$\frac{\varepsilon_{\text{eff}} - \varepsilon_M}{\varepsilon_{\text{eff}} + 2\varepsilon_M} = f\frac{\varepsilon - \varepsilon_M}{\varepsilon + 2\varepsilon_M} \tag{1}$$

must hold. Of course eq.(1) can easily be solved for any quantity of interest, e.g. the dielectric function ε can be calculated if ε_{eff} has been measured and the volume fraction as well as the dielectric function of the matrix ε_M are known.

3.2 Bruggeman

Probably the most frequently used effective-medium theory is the one due to Bruggeman, also known as EMA (Effective Medium Approximation) [8]:

$$f\frac{\varepsilon - \varepsilon_{\text{eff}}}{\varepsilon + 2\varepsilon_{\text{eff}}} + (1-f)\frac{\varepsilon_M - \varepsilon_{\text{eff}}}{\varepsilon_M + 2\varepsilon_{\text{eff}}} = 0. \tag{2}$$

It should be noted that the two materials appear in the Bruggeman formula in a complete symmetric way. Therefor in this model there is no distinction between embedded particles and embedding matrix - in contrast to the Maxwell Garnett approach. In fact, the Bruggeman formula can easily be extended to more than just two components. Then one has

$$\sum_i f_i \frac{\varepsilon_i - \varepsilon_{\text{eff}}}{\varepsilon_i + 2\varepsilon_{\text{eff}}} = 0 \tag{3}$$

where ε_i is the dielectric function of the i'th component and f_i its volume fraction.

3.3 Others

As a representative for the many effective-medium concepts not so well-known as the ones of Maxwell Garnett and Bruggeman the Looyenga formula [9] has been chosen. It has been found to work successful in some cases [10] and will be discussed later a little more. Since the expression for the effective dielectric function

$$\varepsilon_{\text{eff}}^{1/3} = f\varepsilon^{1/3} + (1-f)\varepsilon_M^{1/3} \tag{4}$$

was also derived in a book from Landau and Lifshitz [11] some authors refer to it as the 'LLL' formula [10].

There are many extensions and improved versions of the above stated effective-medium concepts [12, 13, 14]. Usually they extend or modify the topology description and have been proven to be superior in certain cases. A typical example is the extension of the Maxwell Garnett formula from spherical to ellipsoidal particles [15]. Since none of these modified theories has brought a real breakthrough we won't discuss any of them explicitly.

3.4 Performance test

To test the quality of the above given simple effective-medium concepts we now compare their predictions for the optical properties of porous SiO_2-glass to experimental data.

The heterogeneous systems under investigations are taken from a technical process leading from small SiO_2-particles (spherical, typical diameter: 20 ... 50 nm) via chemical and mechanical densification to porous glass, which then in the final step is sintered to compact quartz glass [16]. In the initial highly porous states with large internal surfaces impurities can easily diffuse to the particle boundaries and be removed by a carrier gas. The compact quartz glass obtained after the sinter step is pure enough to be used in optical fibers.

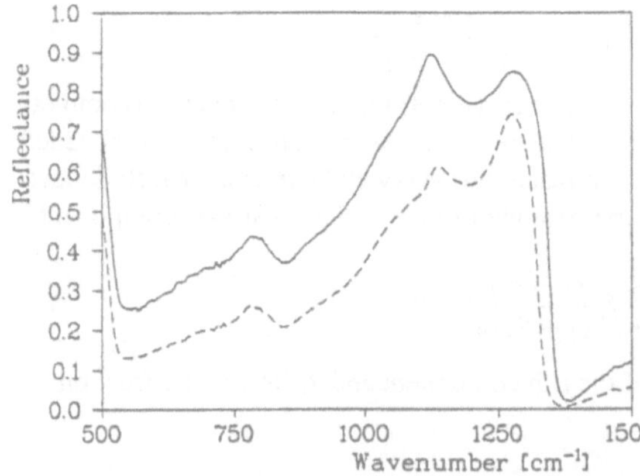

Figure 3
IR reflectance spectra of compact (solid line) and porous quartz glass (dashed) with a volume fraction of glass of 0.47.

To study intermediate sinter states the tempering has been interrupted and the porous glass was examined by infrared reflectance spectroscopy [17, 18]. Fig.3 shows for comparison the reflectance spectra of an intermediate state (SiO_2 volume fraction 0.47) and of the resulting compact glass (incidence angle 70°, s-polarized light).

Clearly the porous glass has a lower reflectance what is expected since its optical density should be lower due to the voids.

Under the assumption that the solid component of the porous system can be described by the 'bulk' dielectric function of compact quartz glass the effective dielectric functions and the reflectance spectra of the porous glass have been calculated according to the simple formulas. Fig.4 shows the results of the Maxwell Garnett and the Bruggeman theory in comparison to the experimentally obtained data. Below 900 cm^{-1} and above 1250 cm^{-1} the agreement is very good. However, in the range in between the theoretical spectra do not fit very well: below 1100 cm^{-1} the Maxwell Garnett formula gives a reflectance too low whereas the Bruggeman result is too large, above 1100 cm^{-1} the situation is reverse. The result of the 'LLL' formula (not shown in fig.4) is similar to that of the Bruggeman theory.

Looking at fig.4 one may ask the following questions: why do the different effective-medium theories give in some spectral regions almost identical results and why do they differ in other? How can the agreement between theory and experiment be improved in the 'unsatisfying' parts of the spectrum? The answers will be given in the next sections where we will discuss the most general expression for effective dielectric functions, the so-called Bergman representation.

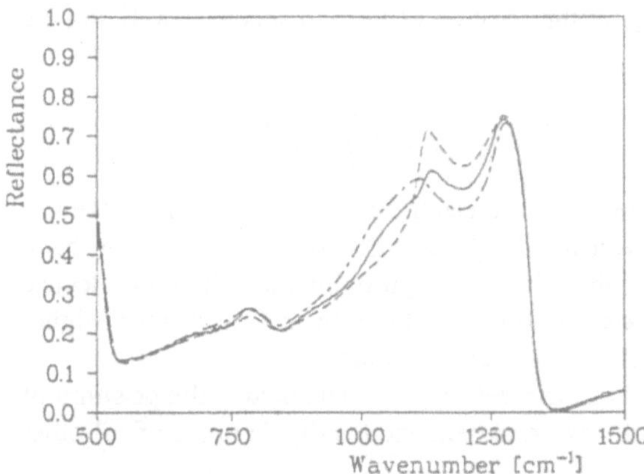

Figure 4
Comparison of the measured reflectance of a porous glass sample (solid line) and theoretical spectra obtained with the Maxwell Garnett (dashed) and the Bruggeman formula (dashed-dotted), respectively.

4 The Bergman representation of effective dielectric functions

In order to find an effective medium theory performing better than the simple concepts shown in the last section it is useful to study what the main features of such a theory should be. This is done in the present section.

A heterogeneous two-phase composite can be considered as a rather complex many-particle system with—at least in dense systems—considerable interaction between the particles. Since the topology and the particle interactions are very complicated in general there is no exact calculation possible. Nevertheless the following simple example shows all the ingredients needed and serves as a motivation for the Bergman representation which then is valid in general.

4.1 Motivation: resonances in a system of coupled dipoles

As an introductory example we show the dielectric response of a few interacting dipoles to an externally applied electric field and discuss the appearance of so-called 'geometrical resonances'. Although this section is not meant to proof anything we show the mathematics quite explicitly in order to demonstrate how the topology influences the total polarization of the system.

We consider an arrangement of N spheres with volumes V_i located at positions $\vec{r}_i$. In order to save space the following notation for the relative positions is used:

$$\vec{r}_{ij} = \vec{r}_i - \vec{r}_j = \begin{pmatrix} x_i - x_j \\ y_i - y_j \\ z_i - z_j \end{pmatrix} = \begin{pmatrix} x_{ij} \\ y_{ij} \\ z_{ij} \end{pmatrix}. \tag{5}$$

When there is an electric field $\vec{E}(\vec{r}_i)$ at the position of the i'th sphere it will show a polarization

$$\vec{P}_i = \varepsilon_0\, \alpha_i \vec{E}(\vec{r}_i) \qquad \text{with} \qquad \alpha_i = \frac{\varepsilon - 1}{\varepsilon + 2} 3V_i. \tag{6}$$

Here we have assumed that the spheres with dielectric function ε are embedded in vacuum which is no severe restriction for the present discussion. In eq.6 we have reduced the sphere to a point dipole which is a poor approximation when the electric field varies significantly over the extension of the sphere. In this case many higher multipole contributions must be taken into account, of course.

If there is a homogeneous external electric field $\vec{E}_0$ the local field at the position of the i'th dipole is a superposition of the external field and the dipole fields of all other dipoles. For the polarization one gets

$$\vec{P}_i = \varepsilon_0\, \alpha_i \left(\vec{E}_0 + \sum_{j \neq i} \vec{E}_j(\vec{r}_i) \right) \tag{7}$$

where $\vec{E}_j(\vec{r}_i)$ denotes the dipole field of dipole j at the position of the i'th dipole:

$$\vec{E}_j(\vec{r}_i) = \frac{1}{4\pi\varepsilon_0} \frac{1}{|r_{ij}|^3} \left(3\vec{P}_j \cdot \vec{r}_{ij} \frac{\vec{r}_{ij}}{|r_{ij}|^2} - \vec{P}_j \right) \quad . \tag{8}$$

Insertion of eq.8 into eq.7 yields a set of coupled equations for the sphere polarizations which can be written in a symmetric form using the quantities $\vec{p}_i = \vec{P}_i / \sqrt{V_i}$:

$$\vec{p}_i = \varepsilon_0 \frac{\varepsilon - 1}{\varepsilon + 2} 3\sqrt{V_i}\vec{E}_0 + \frac{\varepsilon - 1}{\varepsilon + 2} \sum_{j \neq i} \mathcal{W}_{ij} \vec{p}_j. \tag{9}$$

The coupling matrices are given by

$$\mathcal{W}_{ij} = \frac{3}{4\pi |r_{ij}|^5} \sqrt{V_i V_j} \begin{pmatrix} 3x_{ij}^2 - |r_{ij}|^2 & 3x_{ij}y_{ij} & 3x_{ij}z_{ij} \\ 3y_{ij}x_{ij} & 3y_{ij}^2 - |r_{ij}|^2 & 3y_{ij}z_{ij} \\ 3z_{ij}x_{ij} & 3z_{ij}y_{ij} & 3z_{ij}^2 - |r_{ij}|^2 \end{pmatrix}. \tag{10}$$

Finally we obtain a rather compact representation by introducing the 3N dimensional 'super' vectors

$$\vec{P}^{3N} = \begin{pmatrix} \vec{p}_1 \\ \vdots \\ \vec{p}_N \end{pmatrix} \quad \text{and} \quad \vec{E}^{3N} = 3\varepsilon_0 \begin{pmatrix} \sqrt{V_1}\vec{E}_0 \\ \vdots \\ \sqrt{V_N}\vec{E}_0 \end{pmatrix} \tag{11}$$

and the corresponding 3N*3N symmetric 'super' matrix

$$\mathcal{M} = \begin{pmatrix} 0 & \mathcal{W}_{12} & \cdots & \mathcal{W}_{1N} \\ \mathcal{W}_{21} & 0 & \cdots & \mathcal{W}_{2N} \\ \vdots & \vdots & \ddots & \vdots \\ \mathcal{W}_{N1} & \mathcal{W}_{N2} & \cdots & 0 \end{pmatrix} , \tag{12}$$

which then reads

$$\vec{P}^{3N} = \frac{\varepsilon - 1}{\varepsilon + 2} \vec{E}^{3N} + \frac{\varepsilon - 1}{\varepsilon + 2} \mathcal{M} \vec{P}^{3N} . \tag{13}$$

It should be noticed that the matrix $\mathcal{M}$ is symmetric and depends solely on topology information—namely the particle positions and their volumes. Due to its symmetry all the eigenvalues λ_i^{3N} of $\mathcal{M}$ are real. If they and their corresponding eigenvectors $\vec{f}_i^{3N}$ are known eq.13 can be solved for the polarizations of the dipoles in the presence of an external electric field.

In the case of vanishing external field eq.13 reads

$$\frac{\varepsilon + 2}{\varepsilon - 1} \vec{P}^{3N} = \mathcal{M} \vec{P}^{3N} , \tag{14}$$

i.e. there are nontrivial solutions for the polarizations provided $(\varepsilon+2)/(\varepsilon-1)$ is equal to one of the eigenvalues λ_i^{3N} of $\mathcal{M}$. This situation of nonvanishing internal fields in the absence of an external excitation is called a resonance [19] and since the conditions for its appearance depend only on geometrical quantities one can speak of 'geometric resonances'.

Since the eigenvalues of the symmetric matrix $\mathcal{M}$ are real eq.14 tells us that the term $(\varepsilon+2)/(\varepsilon-1)$ must also be a real number in order to see a geometric resonance. For large distances of the interacting dipoles the coupling between them can be neglected and all the matrix elements of $\mathcal{M}$ vanish (see eq.10). In consequence all its eigenvalues must be zero and a resonance is possible only if $\varepsilon = -2$ which is—of course—the well-known dipole resonance. With decreasing particle distances (i.e. larger volume fractions) the degeneracy of the eigenvalues is removed and a splitting is observed in general.

To demonstrate the characteristic features of the interaction effects we consider as an example a system of 19 dipoles (with different sizes) embedded in a vacuum matrix. For three different topologies (which differ significantly with respect to the mean particle distance) the eigenvalues and the corresponding 'resonant' values of the dielectric constant of the dipole material have been computed. Fig.5 shows the results: for large distances the particles behave as noninteracting isolated dipoles (all resonance positions degenerated to $\varepsilon = -2$) whereas for the more dense configurations resonances are found in a rather wide range of ε-values. Note that all the resonances appear for negative real part of ε.

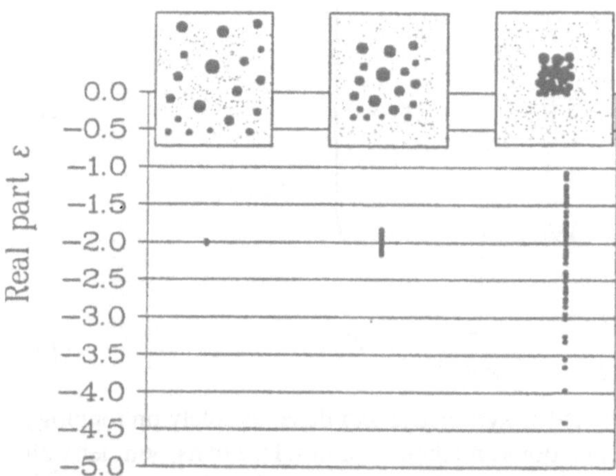

Figure 5
Three dipole configurations and the corresponding values of the real part of the dielectric constant leading to a resonance condition. For large distances (left part) only the resonance of isolated dipoles (i.e. $\varepsilon = -2$) is observed. Here the interaction can be neglected. With decreasing distances (center and right part) the degeneracy of the resonances is removed and a significant splitting of the resonance positions is observed.

To demonstrate the relation to optical spectroscopy we now consider a frequency dependent dielectric function of the dipole material which is shown in fig.6. Such a frequency dependence with negative real part (increasing with frequency) and small imginary part is characteristic for the anomalous dispersion due to strong phonons ('Reststrahlenbande') or free carriers (plasmons). It has been chosen to meet the specifications for the appearance of geometric resonances: one must have an almost vanishing imaginary part and a negative real part.

The frequency dependent total polarization of the dipole system under investigation can then be calculated from eq.13 which can be rewritten as

$$\vec{P}^{3N} = \frac{1}{\frac{\varepsilon-1}{\varepsilon+2} - \mathcal{M}} \vec{E}^{3N} \tag{15}$$

If the eigenvalues and eigenvectors of $\mathcal{M}$ are known, an expansion of both sides of eq.15 in the eigenvectors yields the complete solution of the problem. The total polarization in three dimensional 'real' space can easily be calculated from the solution of the 3N-dimensional 'super' space polarization (see eq.11). Note that for a finite number of dipoles with random, but fixed orientation there will be a slight anisotropic response of the total polarization to the applied electric field. Large random systems with many particles will be statistically isotropic, however.

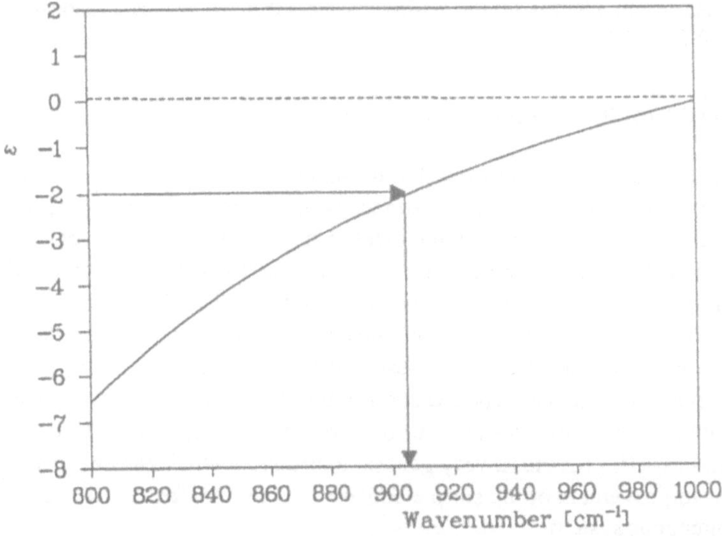

Figure 6 Dielectric function of the model system considered in this section. Note that the real part (solid line) is negative and the imaginary part (dashed) is very small. The arrows mark the dipole resonance position which appears at about 905 cm^{-1} where the real part of the dielectric function is equal to −2.

To demonstrate how geometric resonances show up in the frequency-dependent response of the dipole system we have picked a certain direction in space and show the total polarization in dependence of the mean particle distance in fig.7. Again the resonance of isolated dipoles shows up for large particle distances at around 905 cm^{-1}(where the real part of the dielectric function is almost equal to −2, see fig.6) and splits up into many peaks for increasing 'dipole density', i.e. smaller mean distance.

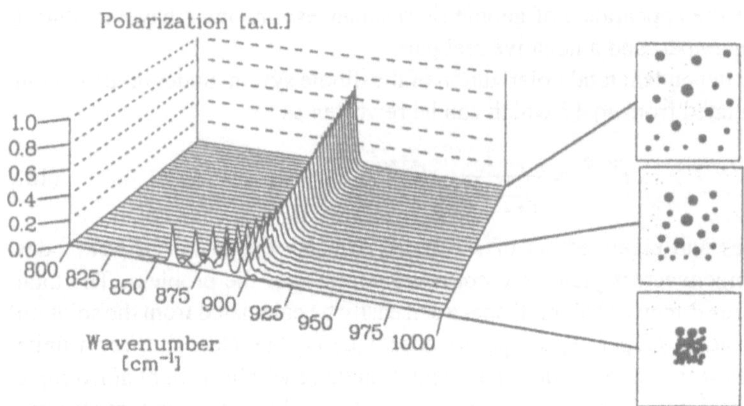

Figure 7 Frequency response of the model system of interacting dipoles in dependence of the 'packing density': for large distances the particles do not interact and only one resonance is observed (back of the graph) whereas for decreasing distances this single sphere resonance splits up into many modes (front).

4.2 Resonance statistics: the spectral density

The last rather mathematical section showed that in many particle systems with considerable interaction complicated geometric resonances occur which do influence the dielectric response of the system to an external electric field significantly. In real heterogeneous systems (without the restriction to dipole interactions and a quite small number of interacting particles) even more resonances can be expected. Of course for more realistic topologies there is nothing to calculate exactly: thousands of irregularly sphaped, partially connected particles would resist very effectively any attempt to determine their geometric resonances. The exact microtopology is not known anyway in general, so a statistical description seems to be a reasonable approach. Here the work of D.Bergman [19, 20] provides a very general expression which can serve as a basis for the desired improvement of the simple concepts to obtain effective dielectric functions of heterogeneous systems.

Bergman considers a two-phase composite with sharp boundaries between the two phases, i.e. the spatial dependent dielectric function $\varepsilon(\vec{r})$ equals either ε_M if there is matrix material at position $\vec{r}$, or ε otherwise. In the quasistatic approximation the complicated local fields that strongly depend on topology obey

$$\vec{D}(\vec{r}) = \varepsilon_0 \varepsilon(\vec{r}) \vec{E}(\vec{r}), \qquad \vec{E}(\vec{r}) = -\vec{\nabla}\Phi(\vec{r}), \qquad div\vec{D}(\vec{r}) = 0 \qquad (16)$$

where Φ is the scalar potential.

On a length scale much larger than the characteristic inhomogeneities the dielectric response is given by the effective dielectric function ε_{eff} which is to be determined by a comparison of the energy density of the true system and that of the replacing effective medium:

$$\frac{1}{2}\varepsilon_0\varepsilon_{\text{eff}}\langle\vec{E}\rangle^2 = \frac{1}{2}\varepsilon_0\frac{1}{V}\int_V \varepsilon(\vec{r})\vec{E}^2(\vec{r})d\vec{r} \qquad \text{with} \qquad \langle\vec{E}\rangle = \frac{1}{V}\int_V \vec{E}(\vec{r})d\vec{r} \tag{17}$$

Here the volume V is much larger than the characteristic inhomogeneity but still smaller than the light wavelength.

Bergman could show that using eq.16 and 17 one can find a represention of the effective dielectric function as a sum of simple poles very much like eq.15 suggests. Here we give an integral formulation which is useful if the system under consideration shows many resonances. With the abbreviation $t = \varepsilon_M/(\varepsilon_M - \varepsilon)$ one has

$$\varepsilon_{\text{eff}} = \varepsilon_M(1 - f\int_0^1 \frac{g(n,f)}{t-n}dn) \tag{18}$$

which is called the 'Bergman representation' for effective dielectric functions. It can be interpreted as follows:

All possible geometric resonances of a two-phase composite occur for real values of the variable t in the interval [0...1]. This corresponds to the condition for particles in vacuum that the dielectric function must be real and negative (compare fig.5). With the integration over n one scans all possible resonance positions in the interval [0...1]. Wether a resonance occurs or not is determined by the 'resonance distribution function' $g(n, f)$—the so-called spectral density— which carries all geometry information. It is remarkable that the spectral density (like the entries in the interaction matrix in the dipole example, see eq.10) depends only on topology. Therefor the Bergman representation clearly distinguishes between the influence of the geometrical quantities (volume fraction f, spectral density $g(n, f)$) and that of the dielectric properties of the constituents (ε_M, t) on the averaging to the effective behaviour. It should be mentioned again that eq.18 holds generally as long as the quasistatic approximation holds. In contrast to the dipole example no further restrictions have been made.

One can show [5] that the zeroth and the first moment of $g(n, f)$ obey

$$\int_0^1 g(n,f)\,dn = 1 \qquad \text{and} \qquad \int_0^1 n\,g(n,f)\,dn = \frac{1-f}{3} \tag{19}$$

where the first condition holds generally but the second one is only true in case of statistical isotropy.

4.3 Parameterizations of spectral densities

Of course the Bergman representation eq.18 is not a great help for concrete problems as long as the spectral density for a system is not known. Very little work has been done up to now to calculate theoretically spectral densities for given microgeometries. Felderhof et al. have treated hard-sphere systems [21, 22], Fuchs has given resonance positions of diluted cubes [35] and suggests to use broad spectral densities for clusters of metallic spheres [36].

Due to this lack of theoretical knowledge our group has fitted spectral densities to experimentally obtained optical spectra of two-phase composites in order to learn which form of spectral densities is required for certain topologies [5, 17, 23, 27, 28, 29, 31, 33]. To do this various parameterizations of the function $g(n, f)$ have been used.

The most flexible one is an intuitive 'graphical' input: $g(n, f)$ is defined by setting some definition points $(n, g(n, f))$ (e.g. with a computer mouse) and doing a spline interpolation in between them. This procedure has the advantage that the number of parameters used (namely the number of points set) is variable and can be adapted to the needs of the problem. In addition the resulting spectral densities are smooth curves nice to look at. Other parameterizations—more suitable for automatic fitting routines— are the sampling of $g(n, f)$ at equidistant positions over the interval $[0..1]$ as sketched in [28] or the use of continued fractions [33].

It has been found to be useful in all parameterizations to use a special parameter to describe the percolation of the system. If the spectral density is splitted according to

$$g(n, f) = g_0(f)\delta(n) + g_{cont}(n, f) \tag{20}$$

then eq. 18 becomes

$$\varepsilon_{\text{eff}} = f g_0(f)\varepsilon + \varepsilon_M(1 - f g_0(f) - f \int_0^1 \frac{g_{cont}(n, f)}{t - n} dn) \, . \tag{21}$$

Considering a system of conducting particles in an insulating host the effective medium inherits the conducting properties by an amount $f g_0(f)$, i.e. $g_0(f)$ is a measure of the 'grade of connectivity' of the particle network and is called 'percolation strength' in the following.

4.4 Performance test

Now since we have parameterization methods with a variable number of parameters we should try what kind of spectral densities give better results as the simple formulas discussed earlier. To do this we consider once again the porous glass sample known from fig.4 and apply the method with spline interpolated definition points set by hand. The comparison of the fit to the measured reflectance spectrum given in fig.8 shows indeed that a significant improvement has been achieved.

The used spectral density (fig.9, solid line) is a rather simple function mainly consisting of two contributions: a broad peak between $n = 0.2$ and 0.3 that is assigned to geometrical resonances of more or less 'isolated particles' and a steep increasing contribution towards $n = 0$ which stands for the resonances of the network building up during sintering. Of course—as we have mentioned before—nothing can be calculated exactly for realistic topologies and therefor the assignment of structures in the spectral density to structures in the microgeometry is highly speculative.

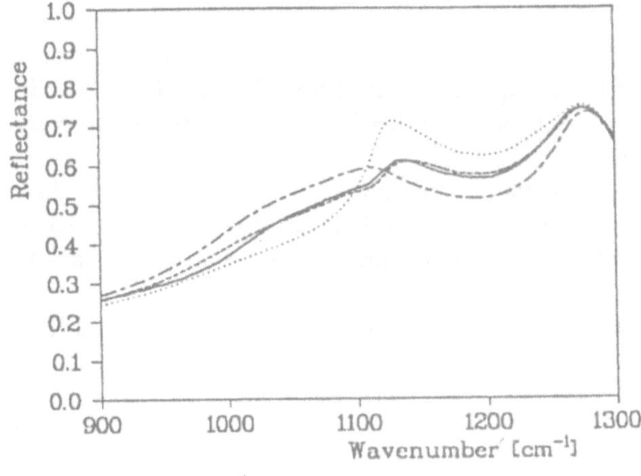

Figure 8
Theoretical reflectance spectra obtained by the Maxwell Garnett (dotted) and Bruggeman formula (dashed-dotted) compared to the result of a spectral density fit (dashes) and the measured data (solid line). The spectral density used is shown in fig.9.

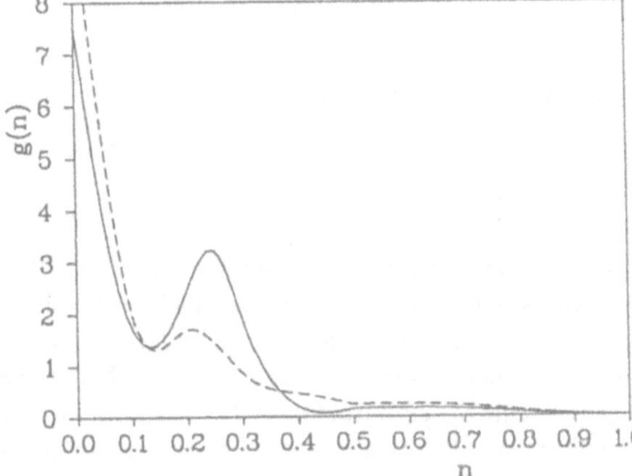

Figure 9
Spectral densities fitted to measured reflectance spectra of two consecutive sinter states. The solid line corresponds to a volume fraction of 0.47 whereas the dashed line belongs to a slightly more dense system ($f = 0.52$).

Fortunately in our porous glass example one can watch how things develop during densification and see if our assumptions are reasonable. To do this fig.9 also shows a fit obtained for a little more compact sample (the volume fraction has increased from $f = 0.47$ to 0.52). If we believe in the above assignments we find that there are less 'isolated particle' resonances and some more network contributions - certainly a result supporting our assumptions.

The success of the use of spectral densities has been shown in many other examples. Since a Bergman representation always can be found as long as the quasistatic approximation holds a flexible parametrization of spectral densities will work properly in any case. Metal-insulator composites have been treated in [23, 29, 31] and [30], mixtures of ionic crystals are described in [27] and [28].

4.5 Transfer of topology information to unknown systems

Of course the determination of spectral densities by a fit to experimental data is only meaningful if the thus obtained knowledge can be used for other problems, too. In general this should be the case since the spectral density depends only on geometry but not on any material property.

Nevertheless one must be careful to choose the right 'training' system which does show clearly the topology influence on the effective dielectric function and hence is suited for a fit. This in turn depends on the dielectric functions of the materials, namely on the combination $t = \varepsilon_M/(\varepsilon_M - \varepsilon)$.

If $|t| > 1$ then the Bergman representation 15 can be expanded in powers of $1/t$

$$\begin{aligned} \varepsilon_{\text{eff}} &= \varepsilon_M(1 - f\int_0^1 \frac{g(n,f)}{t-n}dn) \qquad (22) \\ &= \varepsilon_M(1 - f\frac{1}{t}\int_0^1 g(n,f)\left[1 + n/t + (n/t)^2 + \ldots)\right] dn) \\ &= \varepsilon_M(1 - f\frac{1}{t}\int_0^1 g(n,f)\,dn - f\frac{1}{t^2}\int_0^1 n\,g(n,f)\,dn - \ldots) \\ &= \varepsilon_M(1 - \frac{1}{t}f - \frac{1}{t^2}f(1-f)/3 - \ldots) \end{aligned}$$

where 19 has been used. This result is known as the weak-coupling expansion. If the dielectric functions of the system components are such that t is much larger than 1 then the effective dielectric function will depend only on the volume fraction and no further detail of the microgeometry is obtainable. On the other hand, in such a case all effective medium theories that use the same volume fraction will give the same (and correct) result for ε_{eff} (this also explains why in certain spectral region in the porous glass example all theoretical curves agree, see fig.4).

If on the other hand t gets very close to a real value n_{Test} in the interval $[0, 1]$ then the denominator $t - n \approx n_{Test} - n$ in the Bergman representation 18 will almost vanish and ε_{eff} will very strongly depend on the specific value $g(n_{Test}, f)$. Therefor this would be a system very sensitive to topology effects and different effective medium theories will return quite different values for ε_{eff}. The ideal training system would be one that scans very close along the interval [0...1] on the real t-axis.

Thus looking on the t-function is very instructive. If a spectral density has been obtained looking at a sample with a certain 'topology sensitivity' it may be used for any other case with similar topology and equal or less sensitivity but it must not be used for systems which t-values closer to the interval [0...1] on the real t-axis as the training system. Example for successful transfer of topology information to unknown systems are given in [23, 28].

4.6 Developing new effective medium concepts

If spectral densities for a certain class of topology have been obtained and a particular shape turns out to be dominant then one can think of designing a special effective medium concept tailored for this type of microgeometry. The only thing that has to be done is the choice of a parameterization that generates spectral densities of the correct form with only a few paramters.

For the glass sintering process reflectance spectra of many sinter states have been analyzed and a new effective medium concept has been developed which can describe all measurements fairly good and uses only three parameters. It was suggested by Grosse [30] and reflects a lot of experience with spectral densities. The three parameters control three contributions to the spectral density: one for the network, one peak for the above mentioned 'isolated particle' resonances (splitted and broadened by interaction, see the dipole example) and one very broad 'background' with special shape over the whole 'resonance' interval $[0\ldots1]$. A detailed discussion of this new effective medium theory can be found in [32]. Of course one can try such a new concept even in cases that it is not designed for but a success is not guaranteed, however.

4.7 Reexamination of the simple mixing formulas

Since the Bergman representation for effective dielectric functions holds in general one can ask for the spectral densities contained in the simple effective medium concepts presented above. This could be useful to nderstand which geometrical resonances do play a role in those formulas and hence to select the right formula (or even modify it) for a given problem.

To calculate the spectral density from a given explicit expression for ε_{eff} it is useful to write it in the form

$$\varepsilon_{\text{eff}} = \varepsilon_M(1 - G(t)) \tag{23}$$

which is always possible. Comparing this with the Bergman representation one gets

$$G(t) = f\int_0^1 \frac{g(n,f)}{t-n}dn \tag{24}$$

Now, if the spectral density is to be calculated at n_0 one determines the difference

$$G(n_0 + i\beta) - G(n_0 - i\beta) = f\int_0^1 \frac{-2ig(n,f)\beta}{(n_0 - n)^2 + \beta^2}dn \tag{25}$$

in the limit $\beta \to 0$. Then one can use that

$$\lim_{\beta\to 0} \frac{\beta}{x^2 + \beta^2} = \pi\delta(x) \tag{26}$$

and obtains finally

$$g(n_0) = \lim_{\beta \to 0} \frac{-1}{2\pi i f} \left[G(n_0 + i\beta) - G(n_0 - i\beta) \right] \quad . \tag{27}$$

As an example we show how the spectral density representation of the Maxwell Garnett formula is obtained. From eq.1 one gets

$$\varepsilon_{\text{eff}} = \varepsilon_M (1 - G(t)) = \varepsilon_M \left(1 - \frac{f}{t - (1-f)/3} \right) \quad , \tag{28}$$

i.e.

$$G(t) = \frac{f}{t - (1-f)/3} \tag{29}$$

and

$$G(n_0 + i\beta) - G(n_0 - i\beta) = -2if \frac{\beta}{(n_0 - (1-f)/3)^2 + \beta^2} \quad . \tag{30}$$

Again the term to the right becomes a δ-function in the limit $\beta \to 0$ which leads to the final result

$$g(n_0) = \delta \left(n_0 - (1-f)/3 \right) \tag{31}$$

Here we have just one resonance which shifts with the particle volume fraction (see fig.10). According to the results of the example of interacting dipoles this should be a good description only in the case of very dilute systems. Then, for small volume fractions f, the r particles. A splitting of resonances into a broad distribution for higher volume fractions is not contained in the Maxwell Garnett formula, therefor its use for dense systems is not justified. Also, there is no percolation in this theory even in cases with very high volume fractions (except the trivial case where $f = 1$).It should be mentioned again that extensions of the Maxwell Garnett formula introducing ellipsoidal particles instead of spherical ones remove the degeneracy of resonances partially and a splitting is generated. This is a nice feature which can lead to a considerable improvement in the comparison between measured and theoretical data, but it does not mean at all that the system under investigation must be dominated by ellipsoidal particles. As was shown before the interaction of exact spheres can result in a very similar resonance splitting.

In a similar way as was shown for the Maxwell Garnett formula the spectral densities corresponding to arbitrary effective medium concepts can be determined. Fig.11 shows the result for the Bruggeman formula reflecting some features that have caused the success of this theory. For low volume fractions the spectral density starts with a sharp peak at $n = 1/3$ much like the δ-function of the Maxwell Garnett formula.

But in contrast to a one-resonance behaviour now for increasing volume fraction a broadening into a wide continuum of resonances occurs. For a volume fraction of $f = 1/3$ the continuous distribution reaches $n = 0$ and the term describing percolation—being zero for low values of f—rapidly increases. For volume fractions close to unity the percolation term is dominant over the continuous curve.

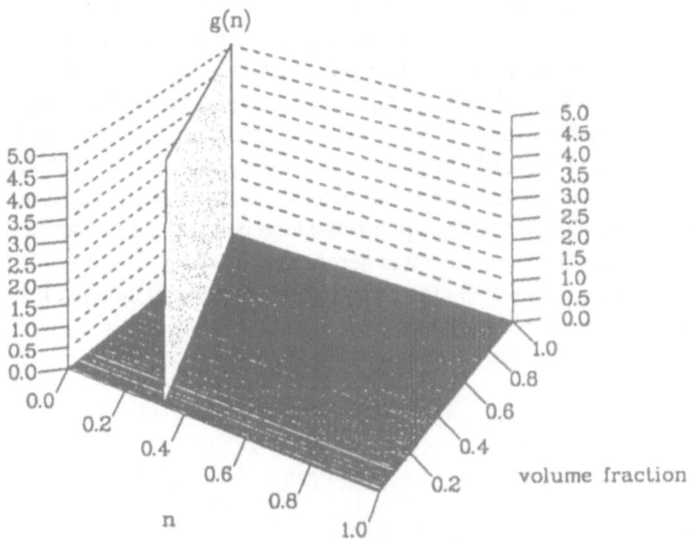

Figure 10 Spectral density corresponding to the Maxwell Garnett formula. As shown in the text it is a δ-function that shifts linearly with the particle volume fraction.

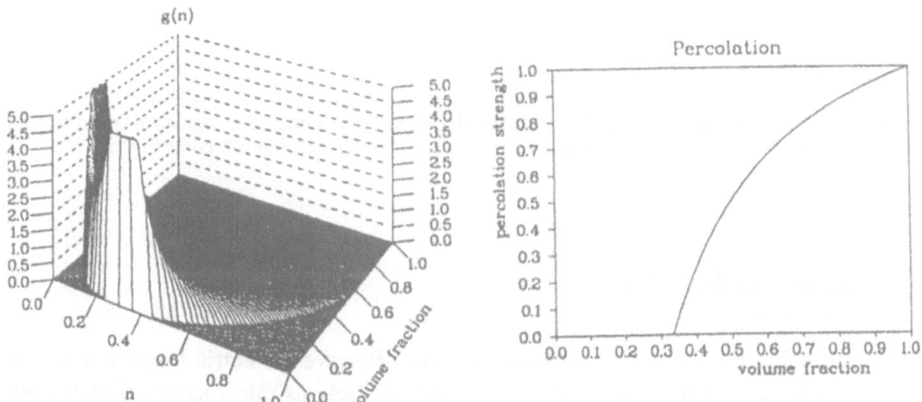

Figure 11 The spectral density belonging to the Bruggeman formula. To the left the continuous part is shown whereas the right graph gives the strength of the percolation term.

The appearance of a percolation threshold is a new quality in comparison to the Maxwell Garnett formula since percolation obviously happens in real ystems above a certain volume fraction. But this advantage should not be overestimated since in many cases two phase composites show an onset of percolation at 'critical' volume fractions very different from $f = 1/3$. Then the Bruggeman formula can be as wrong as the one due to Maxwell Garnett. Especially for systems which are highly connected at volume fractions far below $1/3$ the LLL-formula looks prosperous (see fig.12). It features very broad resonance continua and a percolation term for any volume fraction. Therefor it is the only one of the simple concepts presented here being able to describe highly porous systems where the solid component is percolated at volume fractions of a few percent (e.g. metal blacks, see [29]). On the other hand it should not be used for topologies characterized by unconnected particles.

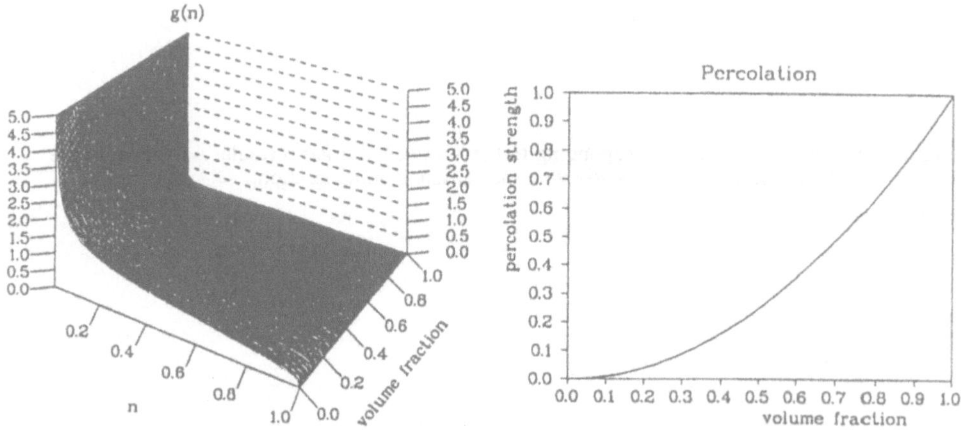

Figure 12 The Looyenga formula represented by its spectral density. For any volume fraction there is a broad continuum of resonances and a nonvanishing percolation contribution.

5 The use of bounds

Although using the Bergman representation for effective dielectric functions means to work with a large number of 'fit' parameters, namely as many as one likes to use to in the parameterization of the function $g(n, f)$, there exist rigorous bounds for the possible values of ε_{eff}. These bounds enclose an area in the complex ε_{eff}-plane which can be calculated (with known dielectric functions ε_M and ε, respectively) according to an algorithm published in [24, 25, 26]. The rigorous bounds are a quantitative measure of the topology sensitivity of a composite and closely related to the behaviour of the t-function discussed before in a qualitative way.

In the case of large differences between the dielectric functions of host and embedded particles the topology influence on the effective dielectric properties will be strong, resulting in a quite large area of possible ε_{eff}-values. In the other extreme of almost equal dielectric functions the bounds enclose a very small spot in the complex ε_{eff}-plane. In this case any spectral density will lead to almost the same effective dielectric function as has been shown already in the discussion of the weak-coupling expansion 22.

For a demonstration we look once again on the porous glass system. Fig.13 shows the bounds for a volume fraction of glass of 0.47 at two frequencies: one (550 cm^{-1}) is chosen from a spectral region very insensitive to geometrical effects (that's where all the different formulas give the same result), the other one (1100 cm^{-1}) is from the region where topology effects are significant. As expected in the first case the bounds enclose a very small area whereas the topology sensitivity of the second case is reflected in a quite large area.

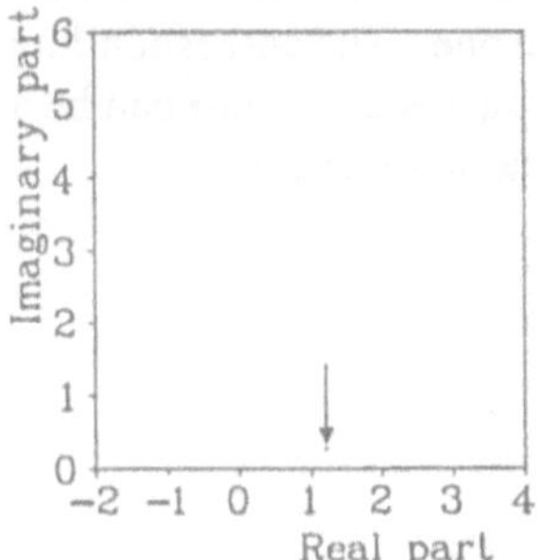

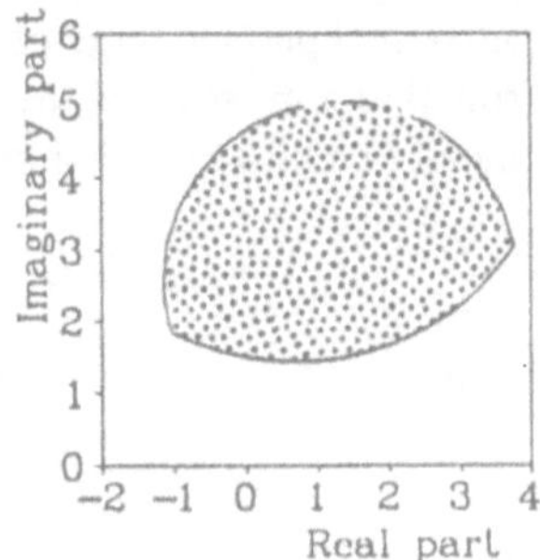

Figure 13 Comparison of the bounds for the effective dielectric function of porous quartz glass (glass volume fraction 0.47) at different frequencies: whereas at 550 cm^{-1} the bounds hardly can be seen (left graph, bounds marked by the arrow) at 1100 cm^{-1} quite different effective dielectric functions are possible.

For the case of reflectance spectroscopy and nontransparent samples (i.e. one is measuring the halfspace reflectance) the bounds can be visualized even more directly: it can be shown that the bounded area in the ε_{eff}-plane can be transformed by conformal mapping into an equivalent area in the complex amplitude reflectivity plane [34]. If one determines the smallest and largest value of the corresponding intensity reflectance (which is what is measured) one can immediately check if the measured value is in that range or not. The size of the possible range of intensity reflectance again is a good measure of 'topology sensitivity'. For the porous glass example fig.14 shows the frequency dependent bounds for the intensity reflectance where the regions of agreement between all theories and the region of discrepancies can be found back very nicely.

An even more extreme case of topology sensitivity is the one of metal particles

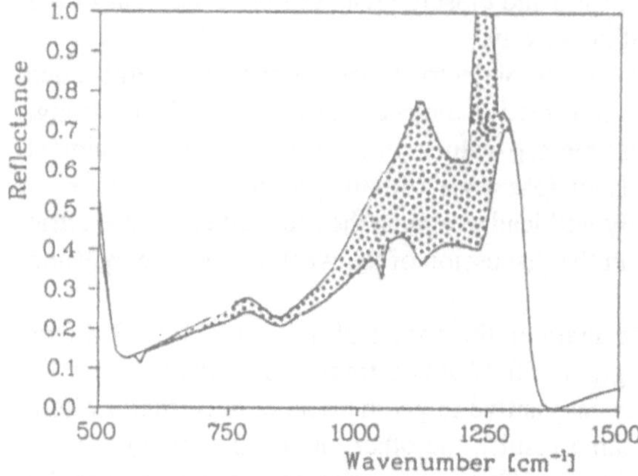

Figure 14
Frequency dependent bounds for the intensity reflectance of porous glass (volume fraction 0.47).

embedded in vacuum. Fig. 15 shows the bounds for a silver-in-vacuum system (volume fraction 0.05). Note the huge range of possible effective dielectric functions (left graph) ranging from large positive real parts to large negative ones. The corresponding intensity reflectance ranges from 0 to almost 1, i.e. the heterogeneous system can be a perfect absorber (known as 'metal blacks') or still show metallic reflectance.

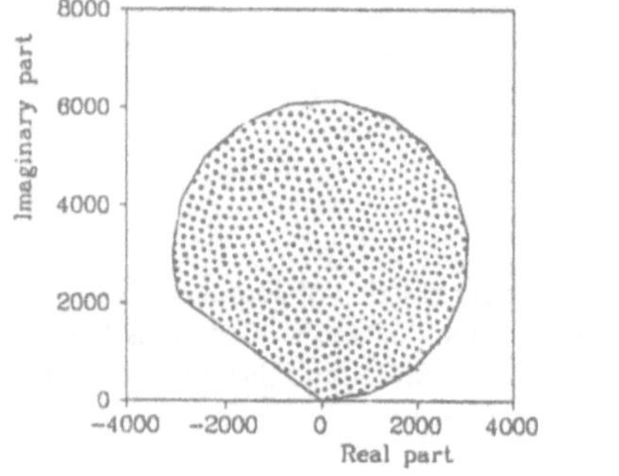

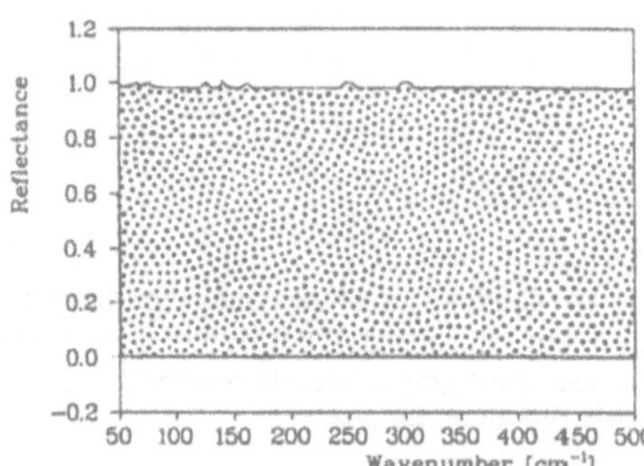

Figure 15 Bounds for a system of silver in vacuum (volume fraction: 0.05): the left graph shows the bounds in the ε_{eff}-plane (for 300 cm^{-1}), the right one the limits of the frequency dependent intensity reflectance.

Searching for size effects in small particles one can make use of the rigorous bounds in the following sense: in order to decide if an observed structure in measured data is due to topological effects or due to a (usually more spectacular!) size effect the area of possible ε_{eff}- values obtainable with bulk dielectric functions should be determined. If the measured data are compatible with effective dielectric functions from inside this area the observation could be explained by topological effects alone. If, on the other hand, the measured data can only be obtained by values clearly outside the bounds the dielectric function definitely differs from that of the bulk.

5.1 A size effect?

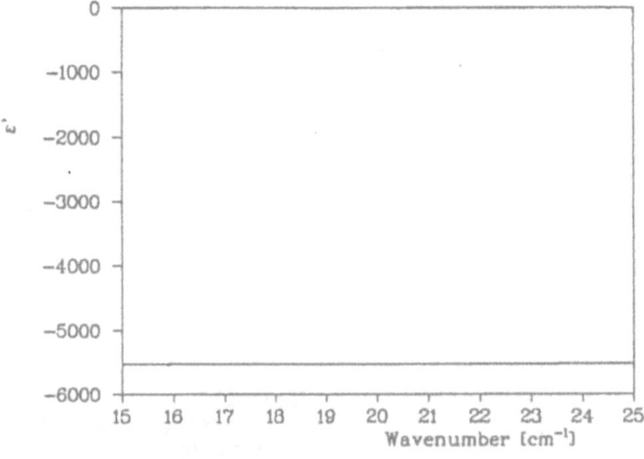

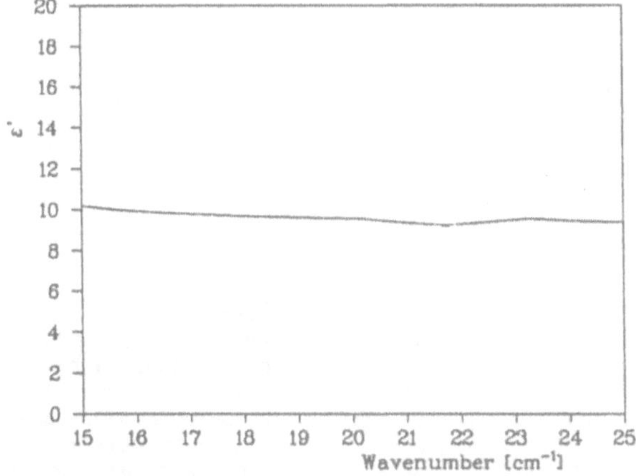

Figure 16 Bulk dielectric functions of Pt (left) and Al_2O_3 (right) which build up the composite discussed in this section. Note that only the real part is shown.

As a warning example we now demonstrate how the use of a wrong effective medium concept (without checking bounds) can lead to very spectacular but not necessary correct results. More details about this example can be found in [31]. We consider a system of platinum particles in an insulating Al_2O_3 host. The dielectric functions of the bulk materials are given in fig.16. The metal volume fraction was 0.08 and micrographs show that the system is not percolated. As a substitute for dc-measurements with the aim to determine the conductivity inside the Pt-particles (and to look for possible size effects) the effective dielectric function was obtained contactless from reflectance and transmittance measurements in the far infrared. It is shown in fig.17.

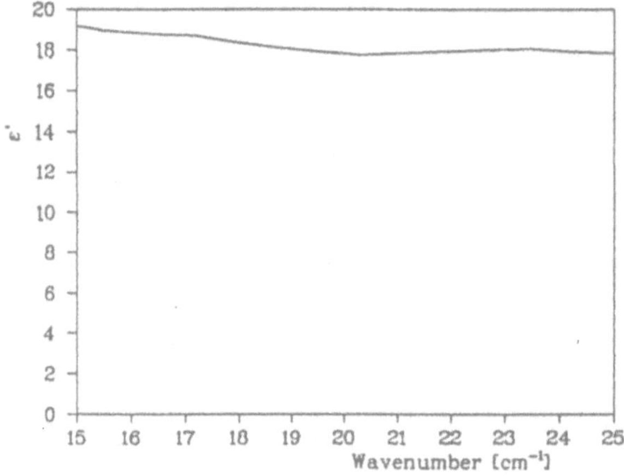

Figure 17
Measured effective dielectric function of the Pt/Al_2O_3-composite (real part).

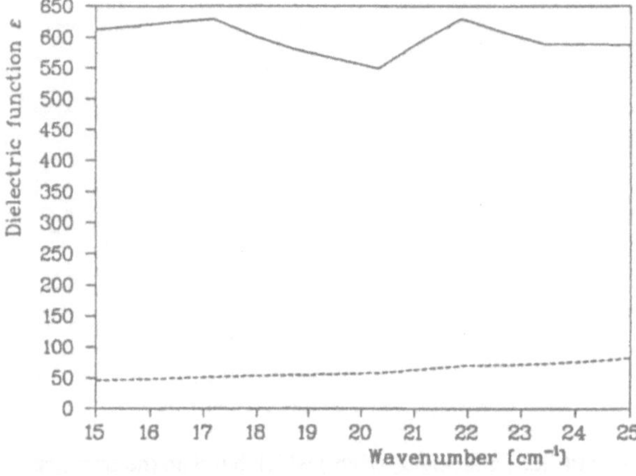

Figure 18
Resulting dielectric function of the Pt-particles in the composite— obtained by using the Looyenga formula. This is the dielectric function of an extremely polarizable insulator (solid line: real part, dashed line: imaginary part).

Now if one assumes a deviation from the bulk behaviour concerning the small metal particles one would like to calculate its dielectric function ε from the measured ε_{eff} (the dielectric function ε_M of the host must be known also, of course). One could be tempted to do this by using one of the simple effective medium formulas which can of course be inverted for the quantity ε. Fig.18 shows the result from an application of the 'LLL' formula (see eq.4) which is very satisfying if one is looking for size effects: the platinum has changed from a good conductor into a very high polarizable insulator.

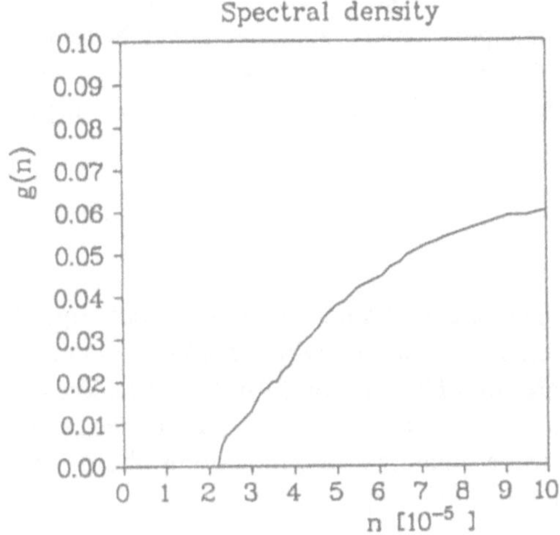

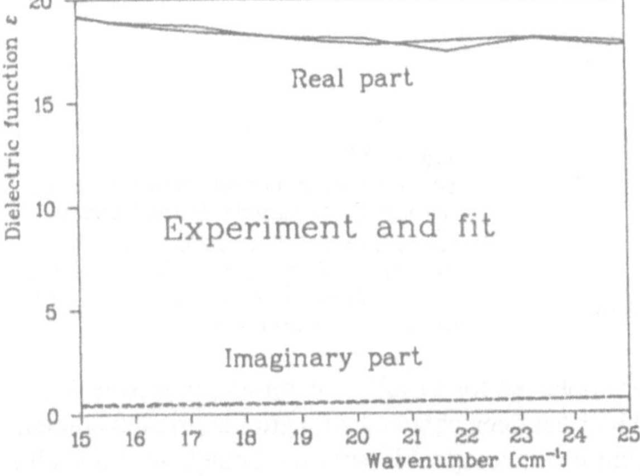

Figure 19
Effective dielectric function of the Pt/Al_2O_3-composite obtained with the bulk dielectric function of platinum compared to the measured one (right). The spectral density used for the fit is shown to the left—it corresponds to an unpercolated topology.

With the knowledge that the spectral density corresponding to the LLL formula contains a percolation contribution (fig.12) the 'spectacular' result can be explained easily: a system of percolated metal particles will always be a conductor—therefor an insulating composite can be compatible with a 'percolating' theory only if the particles are insulators themselves. Since experimentally it was assured that the true system is not percolated the LLL formula must not be applied to this kind of microgeometry and gives unreliable results when one does so. What should be done instead is to look at the bounds obtained with the bulk dielectric function and see if the measured effective dielectric function is within this region. As long as this is the case one cannot distinguish between size and topological effects and a spectral density can be found that reproduces the measured values from the bulk behaviour. Fig.19 shows how this works: measurement and theory are in complete agreement and the spectral density used for the fit looks quite reasonable for an unpercolated system. Note that the metal-insulator composite is very sensitive to the form of the spectral density $g(n, f)$ very close to $n = 0$. None of the simple formulas is flexible enough to give good results in such a case.

5.2 A size effect!

Finally we give an example for the definite detection of a non-topology effect. Fig.20 shows the measured reflectance of a porous silicon sample which has been obtained by electrochemically etching a silicon wafer in a HF solution [40]. It is well known that this kind of process leads to porous silicon structures in the nanometer range with extraordinary properties, first of all to be mentioned a quite strong photo- and electroluminescence in the visible spectral range [38, 39].

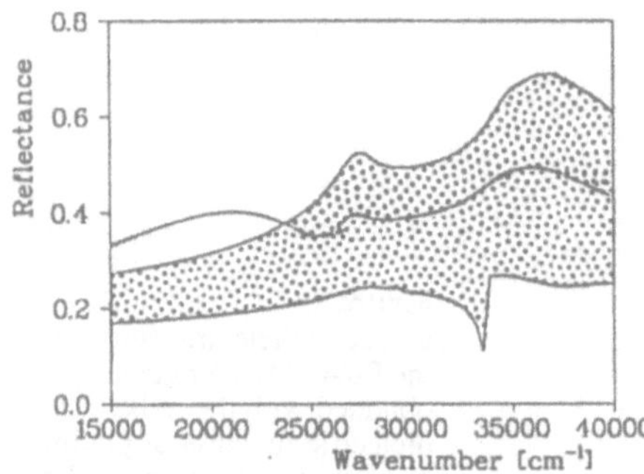

Figure 20
Bounds for the intensity reflectance of a porous silicon sample. In the frequency range from 15000 to 25000 cm^{-1} the experimental spectrum (solid line) clearly is not compatible with bulk values for the dielectric function of silicon.

Fig.20 shows reflectance bounds obtained for a (bulk) silicon-air composite with a silicon volume fraction of 0.75 (which has been obtained with gravimetrical methods). Clearly the porous silicon spectrum is not compatible with the bounds in the visible spectral range (15000 ... 25000 cm^{-1}). Therefor certainly no spectral density can explain this spectrum and changes in the dielectric function of the silicon component must be responsible for the high reflectance around 20000 cm^{-1}. This is—of course—no surprise since the observed luminescence phenomena are very likely due to changed electronic properties with shifted or 'new' gaps.

6 Final remarks

The last example of porous silicon (see fig.20) clearly shows how strong topology effects can be. It underlines again that for nontrivial topologies nontrivial effective medium theories are needed which can be established on the basis of the Bergman representation. Nevertheless, what is needed to do so is a 'training system' (for which everything is known) with similar topology that can be used to obtain the correct spectral density. For 'topology insensitive' cases, however, any effective medium will work properly.

7 Acknoledgements

As you may have noticed I have used the 'we' throughout this article representing the strong interaction between me and my colleagues J.Sturm and S.Henkel. B.U.Felderhof has provided useful external excitations. Finally I would like to thank P.Grosse for taking the risc of employing an 'inhomogeneity' group at his institute - research in this field does not guarantee fast success (nor success at all!). I gratefully thank R.Clasen (Saarbrücken), G.Nimtz (Köln) and H.Münder (Jülich) for placing samples to our disposal.

Bibliography

[1] A.Ishimaru, Wave Propagation and Scattering in Random Media. Vol.1+2, Academic Press, New York and London, 1978

[2] G.Mie, Ann.Phys.25 (1908), 377

[3] M.Kerker, The scattering of light and other electromagnetic radiation. Academic Press, New York, London, 1969

[4] S.Chandrasekhar, Radiative Transfer. Oxford Univ. Press, New York, London, Dover 1960

[5] W.Theiß, Doctoral Thesis, RWTH Aachen 1990

[6] M.P. van Albada, A.Lagendijk, Phys. Rev. Lett. 55 (1985), 2692

[7] J.C.Maxwell Garnett, Philos.Trans.R.Soc.London 203 (1904), 385

[8] D.A.G. Bruggeman, Ann.Phys.(Leipzig) 24 (1935), 636

[9] H.Looyenga, Physica 31 (1965), 401

[10] P.Marquardt, G.Nimtz, Phys. Rev. B 40, No.11 (1989), 7996

[11] Landau-Lifschitz, Lehrbuch der theoretischen Physik, Band VIII: Elektrodynamik der Kontinua. Berlin: Akademie-Verlag 1967, 55

[12] P.Sheng, Phys. Rev. Lett. 45, No. 1 (1980), 60

[13] A.Wachniewski, H.B.McClung, Phys. Rev. B 33, No.12 (1986), 8053

[14] J.Lafait, S.Berthier, L.E.Regalado, SPIE Vol. 652 (1986), 184

[15] R.W.Cohen, G.D.Cody, M.D.Coutts, B.Abeles, Phys. Rev. B 8 (1973), 3689

[16] R.Clasen, Habilitationsschrift, RWTH Aachen, 1991

[17] M.Hornfeck, W.Theiß, R.Clasen, J. of Noncryst. Solids 145 (1992), 154

[18] R.Clasen, M.Hornfeck, W.Theiß, SPIE Vol. 1513 (1991), 243

[19] D.Bergman, Bulk physical properties of composite media. Les methodes de l'homogeneisation, Edition Eyrolles (1985)

[20] D.Bergman, Physics Reports C 43 (1978), 377

[21] K.Hinsen, B.U.Felderhof, J. Chem. Phys. 94 (1991), 5655

[22] B.Cichocki, B.U.Felderhof, J. Chem. Phys. 90 (1989), 4960

[23] M.Evenschor,P.Grosse, W.Theiß, Vibrational Spectroscopy 1 (1990), 173

[24] G.W.Milton, J. Appl. Phys. 52 (1981), No.8, 5286

[25] G.W.Milton, J. Appl. Phys. 52 (1981), No.8, 5294

[26] B.U.Felderhof, Physica 126A (1984), 430

[27] J.Sturm, P.Grosse, W.Theiß, Z. Phys. D 20 (1991), 341

[28] J.Sturm, P.Grosse, W.Theiß, Z. Phys. B 83 (1991), 361

[29] T.Eickhoff, P.Grosse, S.Henkel, W.Theiß, Z. Phys. B 88 (1992), 17

[30] P.Grosse in Festkörperprobleme / Advances in Solid State Physics, Vol. 31, ed. by U.Rössler (Vieweg, Braunschweig 1991), 77

[31] J.Sturm, P.Grosse, S.Morley, W.Theiß, Z. Phys. D, to be published

[32] J.Sturm, Doctoral Thesis, RWTH Aachen 1993

[33] E.Gorges, P.Grosse, J.Sturm, W.Theiß, in preparation

[34] E.Gorges, private communication

[35] R.Fuchs, Phys. Rev. B 11, No.4 (1975), 1732

[36] F.Claro, R.Fuchs, Phys. Rev. B 33, No.12 (1986), 7956

[37] P.Grosse, private communication

[38] L.T.Canham, Appl.Phys.Lett. 57 (1990), 1046

[39] V.Lehmann, U.Gösele, Appl. Phys. Lett. 58 (1991), 856

[40] W.Theiß, P.Grosse, H.Münder, H.Lüth, R.Herino, M.Ligeon, Proc. of the MRS fall meeting in Boston (1992), Symposium F, in press

Systems on Chips: The Microelectronics Challenge of the Next 20 Years

Dr. Armin W. Wieder, Siemens AG, Munich

I.Physik. Inst. RWTH Aachen, Postfach o. Nr., 52056 Aachen

Summary: Even if the present pace of progress continues, the prospects for microelectronics beyond the year 2000 will not be limited by technological constraints.

Equally positive statements can be made about future developments in applications. Thus, the computational power of a chip can be boosted by about four orders of magnitude to 10^5 - 10^6 MIPS on the basis of verified structures. However, this implies that heed be paid to the various recommendations given in this paper as regards technology, circuit design, layout and architecture.

Conventional stand-alone markets for memory and logic ICs are predicted to grow by 10-15% annually for the next decade. But user-specific memory as well as logic with on-chip memory (volatile and non-volatile) will become increasingly important. These novel "systems on chips" will eliminate the bottleneck represented by the on-chip/off-chip metallization. Consequently, the technical approximation to "biological systems" will become feasible and lead to an increase in system performance of several orders of magnitude. These on-chip systems contain today's system know-how and will become the modules of the hypersystems of tomorrow.

1 Introduction

The historical groundwork for microelectronics, which has now changed our lives so dramatically, was laid down in the recent past. Schockley, Bardeen and Brattain [1;2] discovered transistor action at Bell Laboratories in 1948. About 12 years later, Kilby [3] and Noyce [4], from Texas Instruments and Intel respectively, devised planar technology and the simultaneous manufacture of transistors and wiring systems and thus invented integrated circuits. This then triggered the ensuing rapid development in microelectronics. In 1970, bipolar circuits with around 10^3 components (transistors, diodes, capacitors, and resistors) and switching times of some hundreds of nanoseconds could be produced. At the same time, the problem of interface charges at the oxide-silicon boundary was solved by suitable annealing treatment and the first MOS transistors could be successfully produced. Then in 1980, bipolar circuits with 10^4 transistors and 200 nanosecond delay time, and MOS circuits with 10^5 devices were obtained on an industrial scale. Today (1992), this technique makes available bipolar circuits with more than 10^5 devices and MOS circuits with almost 50 million components for applications in communications, data processing, automation, traffic technology, medicine and many other sectors of great importance in our lives [5].

Since about 1960, the semiconductor components market has been growing at a compound annual growth rate of approximately 15%. Forecasts assume that future growth rates will be the same. The result will be that the IC world market, amounting to DM 250-300 billion in the year 2000, will be about as large as the market of the automobile industry worldwide at the same time. It should be noted that this IC market is a pure components market whose technology push has a multiple effect on the systems business sector because without these key components, many systems cannot be realized at all or at least not in this form. The predictions made by the Club of Rome have therefore been vindicated, namely calling integrated circuits "the modern raw material of the industrial nations" whose importance for the future can hardly be overestimated [5].

To assess the development potential of microelectronics, we must look at possible "show-stoppers" in the physical, technological or business sectors that could well put the brakes onto the extrapolated development (Fig. 1).

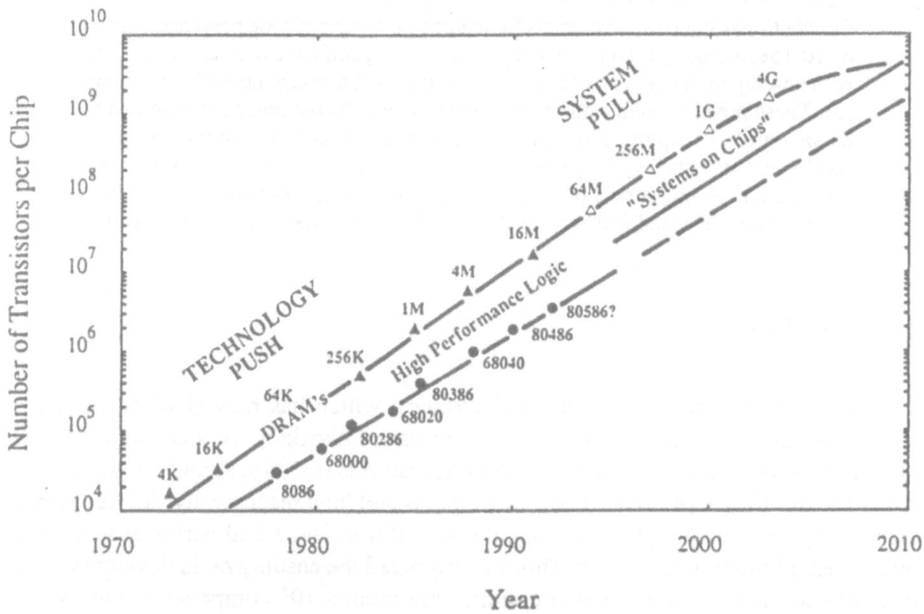

Figure 1 Development of chip complexity for memory (Moore curve) and logic devices

2 Technological perspectives

The present state of microelectronics is best reflected by the various generations of dynamic memory (DRAM = Dynamic Random Access Memory). Today the 16M DRAM with approximately 50 million devices and a 0.6 μm minimal feature size is in the preproduction stage at leading international semiconductor manufacturers. Current developments are represented by the 64M DRAM with a 0.4 μm minimal feature size, that is at present in the evaluation phase. Research work is under way in industry which would already allow the production of 256M DRAMs with around 500 million components per chip in the nineties (solution as shown in Fig. 2).

Figure 2 REM photograph [6] of Si towers 2 μm in height and approx. 0.3 μm wide. Addressing of 256 M cells takes place on the top of the towers whereas storage is on the side walls.

To assess this development potential, a basic plausibility evaluation is helpful in showing that today's transistor structures down to the critical lateral dimensions of 0.05 μm can function even when subject to statistical fluctuations with sufficiently small (approx. 10%) tolerances (Fig. 3). Explorative experiments (Fig. 4) have now confirmed the feasibility of such transistors. The accompanying intensive simulation work has at the same time expanded technological know-how to such an extent that ICs with critical dimensions of 0.05 μm, i.e. with almost 10^{10} transistors per cm^2 (equivalent to about 1000 4-Mbit DRAMS per cm^2) can now be manufactured with certainty. Even if no breakthrough on a quantum-leap scale takes place in the device sector in the next 20 years, an unlikely scenario, the technology at least will offer no obstacles to an evolutionary development with the same growth rate as in the past (Fig. 1).

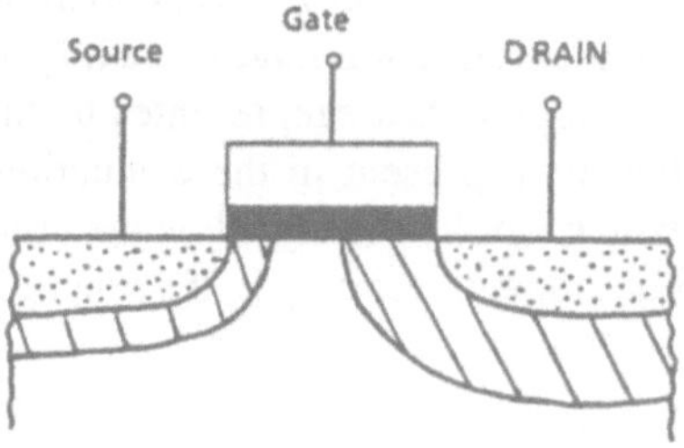

1. $t_{ox\,Min}$(direct tunneling) ≤ 30 Å

2. E_{Max}(breakdown, dielectric) ≈ 10 MV/cm

   $\Longrightarrow N_{A\,Max}$(only just invertable) $\approx 10^{18}$ cm^{-3}

3. $X_{RLZ}(1V, N_{A\,Max}) \approx 0.03\ \mu$m

   $\Longrightarrow L_{G\,Min} \approx 0.05\ \mu$m; $A_{Tr\,Min} \approx 8F^2_{Min} \approx 10^{-10}$ cm^{-2}

4. Tolerance Analysis, Fluctuation

   N^{Si}_{Atoms}(minimal transistor) $= 0.05\ \mu m^3 \cdot N_{Si} \approx 10^7$Si atoms

   N^{Dopant}_{Atoms}(minimal transistor) $= 0.05\ \mu m^3 \cdot N_{A\,Max} \approx 10^2$dopant atoms

   Fluctuation $\sqrt{N} \approx 10$ % (acceptable)

Figure 3 Plausibility evaluation of 0.05 μm transistor structures

Besides the economic motivation, the evolutionary development and the fractal structure of microelectronics can be seen to be the basic principles of a continuous advance. Advances in diverse disciplines (materials science, lithography, process architecture (transistors etc.) analysis, circuit design, system architecture, CAD, simulation, ...) take place in small "quantum leaps" which then become effective to various degrees depending on their compatibility with the evolutionary progress in microelectronics. Progress and system friendliness play a decisive role in the generation change that is leading to microelectronics itself creating "intelligence" in the form of the increasingly powerful computers which are required for its own development.

Far beyond the developments foreseeable today, device innovations (quantum coupled devices) are already in the offing that could push microelectronics to the next evolutionary level. These "device quantum leaps" use, in addition to the well-known physics of the pn-junction and MOS systems, principles such as heterojunctions and quantum mechanical interactions that allow both the amplifying and storage functions to be implemented in a single component. The first components of this type are shown in Fig. 5. They could gain increasingly importance against conventional transistors, starting with the year 2000 onwards.

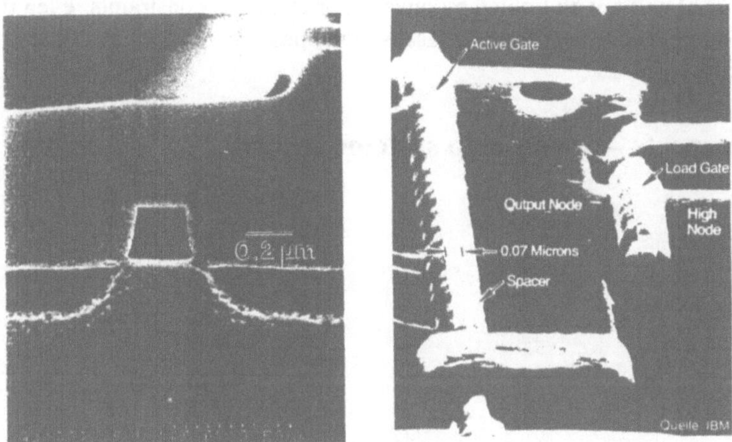

Figure 4 Experimental verification of minimal transistors: 0.2 μm transistors [8] and 0.07 μm transistors [9] operating at 77 K and 300 K

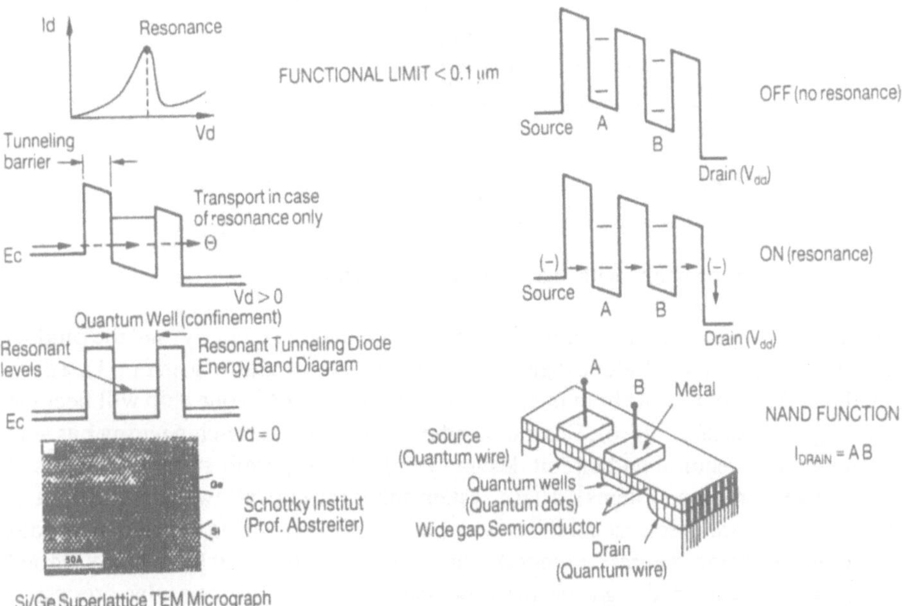

Figure 5 Direct implementation of a NAND function by means of quantum effect devices

In summary, it can be said that the technological perspectives for microelectronics beyond the year 2020 will not be limited by physical or technical constraints, even if the pace of development continues to be as fast as in the past.

3 Future prospects – are on-chip systems realistic?

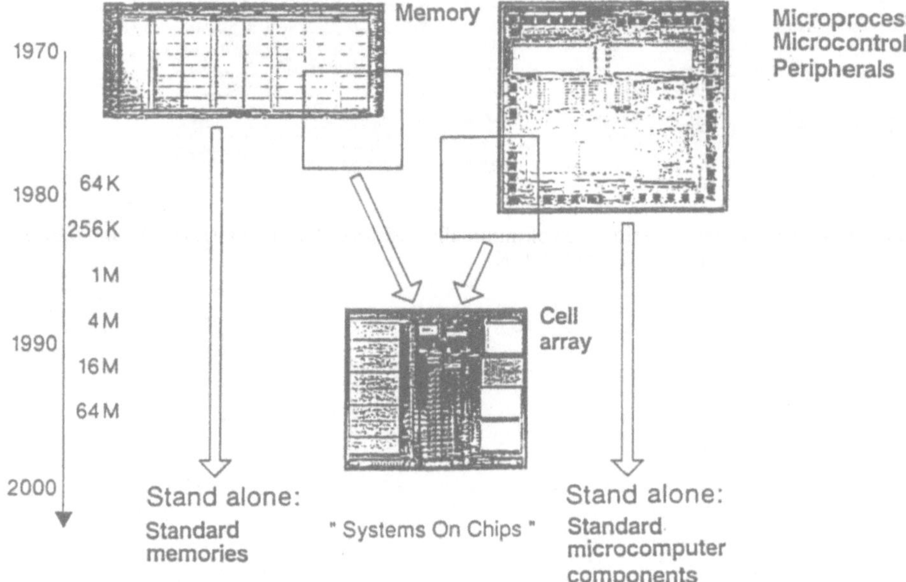

Figure 6 "Systems on chips" with logic and memory functions

In addition to the product lines for memory chips (stand-alone memory) and the product lines for logic chips (stand-alone logic) with their predicted trends (e.g. in Fig. 1), a class of modules that incorporate both memory and logic functions in one chip will become increasingly important (Fig. 6). Eliminating the restriction of inter-chip wiring has lead to dramatic architectural and circuit design synergies that permit system solutions at the chip level (on-chip systems). These system solutions now allow the realization of completely new architectural concepts with extremely high wiring density analogous to biological systems (neural networks). This type of development has already started using memory-supported logic on the processor side (RISC, superscalar processor, single chip computer etc) as well as logic-supported memory (video RAM, feature box etc) (Figs 7, 8).

The architectural and circuit design benefits due principally to the massive parallelization in both the space domain (hardware complexity) and the time domain

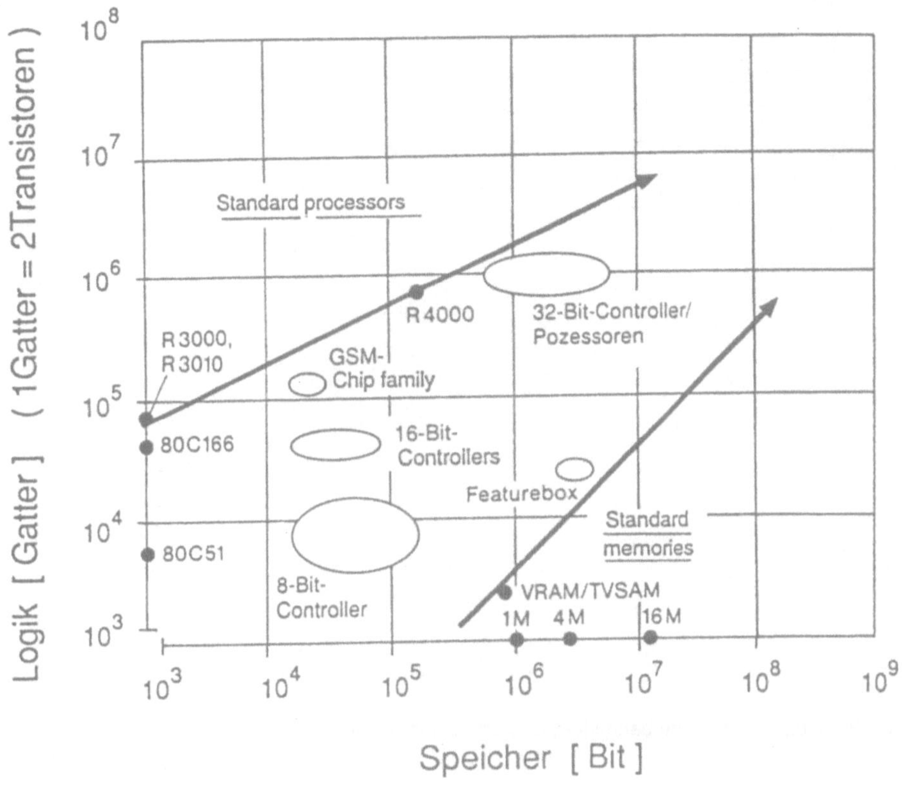

Figure 7 "Systems on chips"

(multiplexing) become most clearly apparent in the performance versus time curve for computers (Fig. 9). The classical technology-optimized development line that will break through the gigaflop threshold in the year 2000 is supported by a highly parallel and modular computer architecture that will use networked transputers to attain computational performance of the teraflop level by the year 2000 for an investment of around 20 million dollars.

An important consequence of the massive parallelization in the space and time domains is a dramatic increase in the transmission rate requirement of components/mod"-ules for use in networks. Empirical relationships of the kind shown in Fig. 10 can be established for chip systems for applications in computers and communications [11]. The computational performance and transmission rate of modern systems are coupled. Thus the requirement is for a transmission rate of 1 x Mbit/s (Ruge) and a storage capacity of 1 x Mbit (Amdahl) per 1 x MIPS computational performance. One of the consequences of this for future processors with 1000 MIPS and above is the increasing importance of fast static memory (SRAMs) or rather the integration of fast

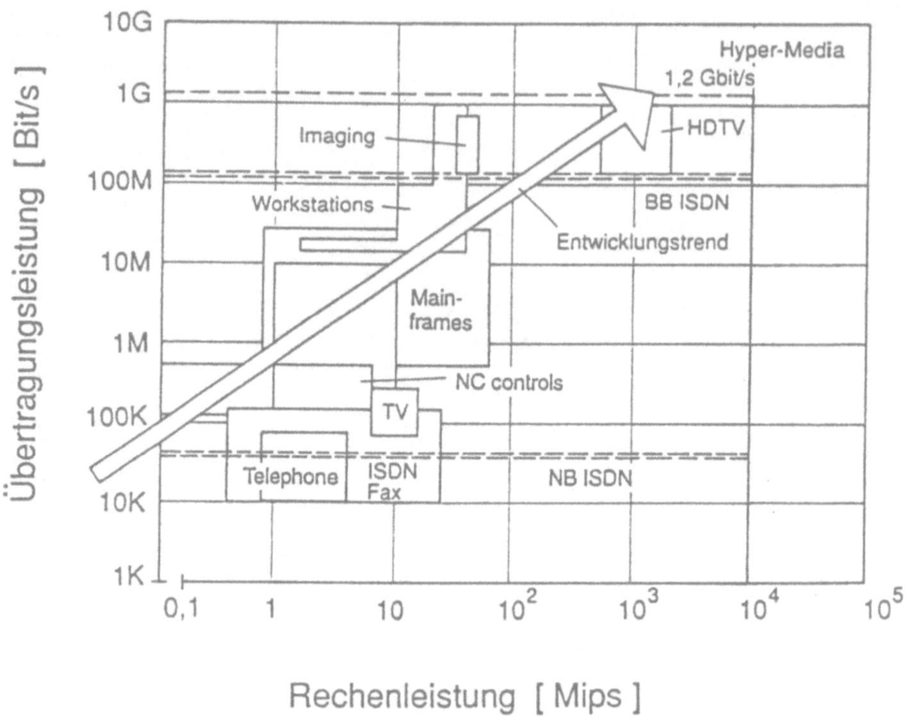

Figure 8 Requirements on bandwidth and processor performance

microprocessors and high-speed SRAMs or non-volatile memories (EEPROMs) on one chip.

Looking to the year 2000 and beyond, applications for artificial intelligence disciplines assume performances far in excess of these figures. Professional expert systems, voice recognition systems for real-time applications with a vocabulary of 10^4 words or pattern recognition systems for real-time operation in HDTV quality call for an increase in performance from today's 100-500 MIPS to 10^5-10^6 MIPS over the next 10 years. This represents a surge of 3 to 4 orders of magnitude which cannot be attained primarily through an evolutionary improvement of technological performance alone. Novel architectural solutions with massive time and space parallelization certainly have the necessary potential for such quantum leaps in performance [12]. The precondition for these various architectural solutions is a massive increase in chip complexity, again about 4-5 orders of magnitude (Fig. 11).

In this respect, CMOS technology stays the dominant IC technology since it has unique advantages as regards high integration. These are simple design (can be automated), signal to noise (external, internal, technical noise), insensitivity to supply changes, low quiescent power dissipation and high chip yield through its technological

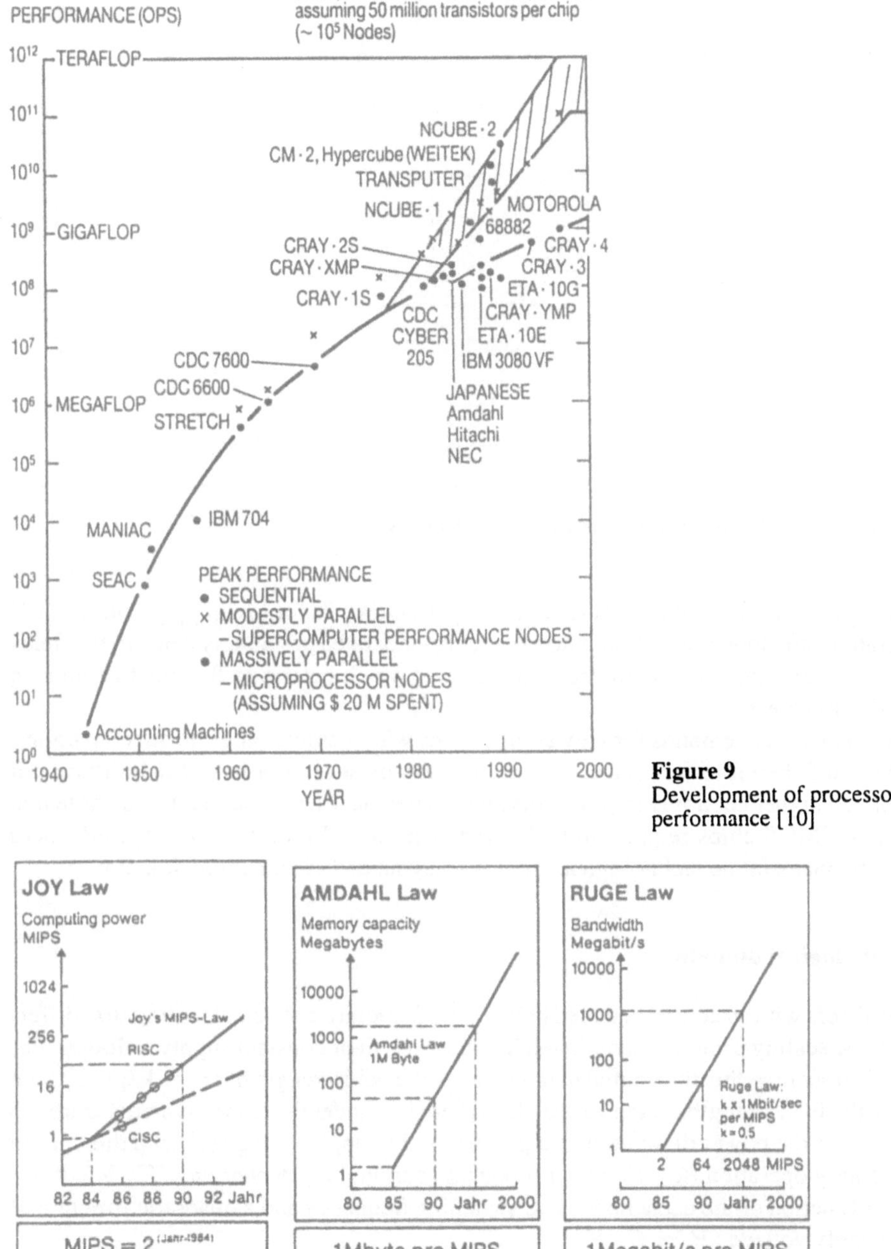

Figure 9
Development of processor performance [10]

Figure 10 Empirical relationships [11]

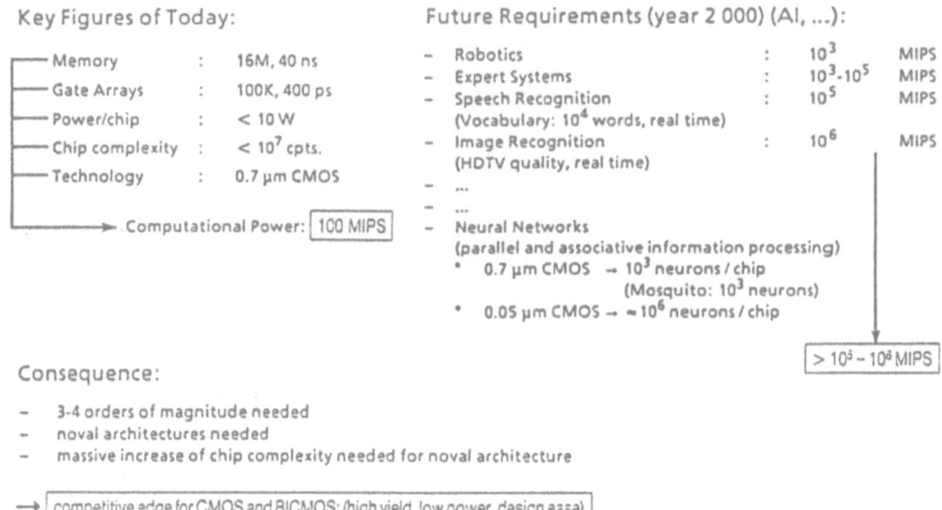

Figure 11 CMOS/BICMOS – the winning technology

leading role. This development is boosted by the demands for low supply voltage (1 V operation) and low power dissipation for battery-operated portable systems in the fields of computers and information/communications which belong to the most promising application areas.

However, the demands for new architectures with massive time and space parallelization and therefore for greater chip complexity by several orders of magnitude call for much higher economy in power dissipation per electrical function. Let us note that today's CMOS chips require up to 10 watts per cm^2. With a view to this, advances must be made in the technological, circuitry, layout and architectural domains.

Technological domain

Transistors with structures reduced below 0.5 μm require a physical scaling that differs from the scaling principles employed hitherto with their constant supply voltages. For physical scaling, the differential parameter, i.e. the field strength has to be kept constant since the field strengths at the gate dielectric are just under the respective breakthrough values. As the power dissipation is a product of the supply voltage (V) and the current per binary operation ($Q \cdot f$) which produces charging and decharging ($C \cdot V$), it is a linear function of the capacitance and switching frequency and a quadratic function of the supply voltage ($P_d = C \cdot V^2 \cdot f$).

On decreasing the structures below 0.5 μm, the increase in the clock frequency will be merely linear due to velocity saturation of mobile carriers in the channel. The clock frequency could even remain constant if the effect capacitanties of stray become more significant in the future. As the structure size becomes reduced still further, the integral chip capacitance per unit surface will increase approximately linearly because of the greater edge and coupling capacitances. On the other hand, the supply voltage is reduced by a power of two through physical scaling so that technological factors will make the power dissipation per unit surface will independent of structural reduction and limited in the worst case to five times today's value (approx. 10 W/cm^2) of a 0.8 μm CMOS technology (Figs 12, 13).

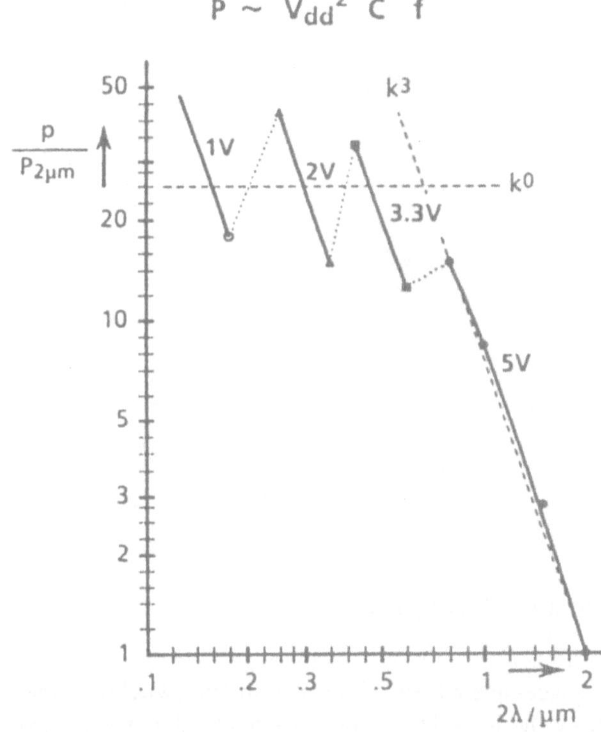

Figure 12
Power dissipation per unit surface will increase at most by a factor of five of today's value [13]

Circuitry domain

To achieve minimal power dissipation on the circuit design level, it is recommended to use static CMOS technology. Dynamic CMOS technologies appear at first sight to be more advantageous since fewer transistor gates need to reverse their charges. However, the switching statistics of such technologies clearly entails a drawback. For example, NORA circuit nodes that are in logic state 0 are nevertheless charged and discharged at every clock. Another circuitry aspect is the effect of glitches on the statistics of specific gate switching. There are critical paths which have different delay

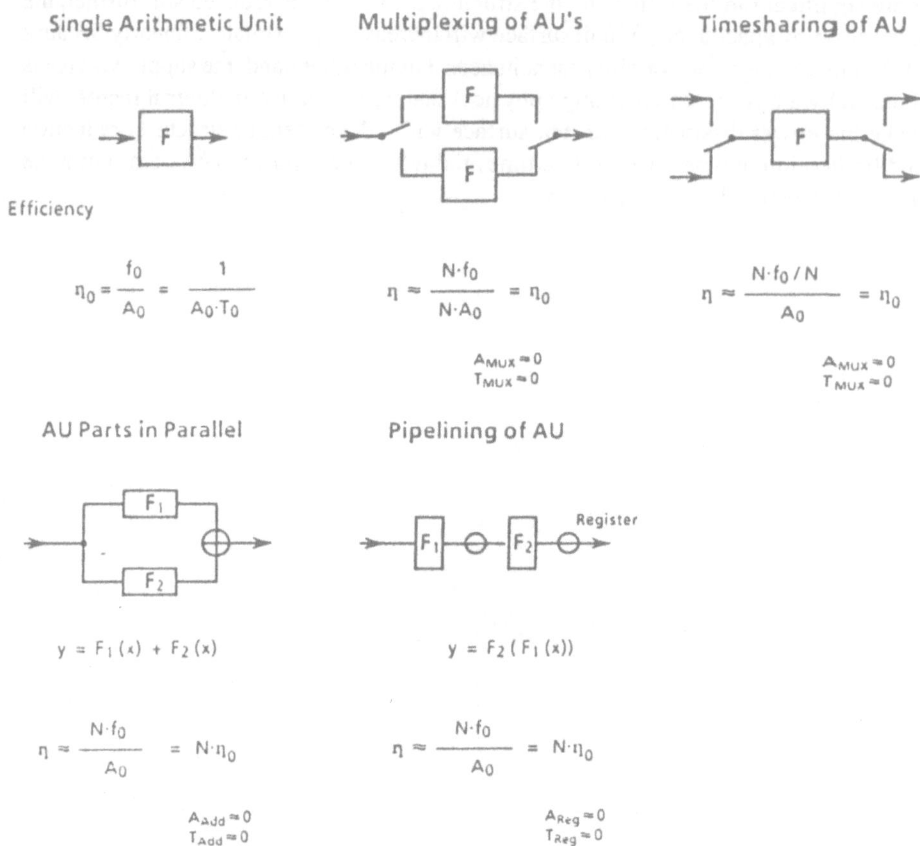

Figure 13 Efficiency of various architectural concepts [13]

times, especially in the case of successive adders with carry, so that switchings may occur at the internal nodes before the logic level attains its final value. Carry-save techniques and/or pipelining reduce the probability of glitches [13].

Further ways of reducing the power dissipation at the circuitry level mainly lie in cutting down on the switching activity or on the transport of data. For example, long arrays of shift registers should be abandoned in favor of DRAMs with pointer control.

Furthermore, innovative solutions including optical clock systems could become expedient for reducing power dissipation at the circuitry level (up to 50% for high performance circuits).

At the layout/design level, concepts that result in lower wiring capacitances are to be favored. These are generally design concepts with considerable local interconnect such as pipelining architecture and few "linking networks" which require to be operated with a low logic level span.

Architectural domain

To compare the various architectures, it is necessary to define an efficiency factor n that takes into account at a constant functionality F the required silicon surface area A, the sampling frequency f_s, depending throughput rate as well as the resulting specific power dissipation $P \cdot f_s$ [13]:

$$n = \frac{F \cdot f_s^2}{A \cdot P}$$

Various ways of parallelization and mutiplexing are considered in the following with regard to their efficiency and compared for a simple arithmetic unit with the functionality F as shown Fig. 13. On the assumption that the hardware requirement for multiplexers and demultiplexers is negligible, the efficiency cannot be raised either by a parallel configuration (Fig. 13b) or by multiplexing (Fig. 13c). The ratio of throughput to required area as well as the specific power loss remain constant in both cases. If the operation F were to be split into suboperations which could be performed by parallel arithmetic subunits (Fig. 13d), the efficiency would at best increase with the number of these parallel subunits. However, the assumption that the hardware requirement for the summation of the partial results is negligible, is difficult to satisfy. On the other hand, the efficiency increases when space and time parallelization is employed in the pipeline technique (Fig. 13e) because the amount of hardware for intermediate storage can be kept low. The assumption here is that the operation to be performed can be split up into a sequence of suboperations F.

The limitations on the application side can be pinpointed with some ease, inasmuch as most of the key applications, e.g. in information technology, communications, data processing and consumer electronics, can be represented by largely isolated methods of solution, pipelining or by combinations of both.

Harmonization

When it comes to integral optimization, harmonization of the different measures taken at the different levels, or rather the avoidance of conflicts between them, is of primary importance. A detailed examination reveals that the various recommendations and physical scaling that are required on technological grounds mutually support each other so that the whole package of measures can be regarded as leading to harmonization without any contradictions. Recommendations at the layout and design level are for simplicity of design, local interconnect and simplified automation. Recommendations at the circuitry level are for static CMOS and save techniques, and those at the architectural level for parallel and pipelining structures with a high degree of modularity and locality. Thus, chip performance can be improved by doubling the hardware and increasing the clock frequency.

The conclusion is that, with due regard to the measures described chip systems with 10^{10} -10^{11} devices and a performance of 10^5 -10^6 MIPS can be produced. The necessary methods and structures have already been verified, at least in principle and at an extremely low level of maturity. Improvement in efficiency has an improvement potential of about 400, the chip size of 2 to 5 and the clock frequency one of 5 to 10. However, these performance data must be corrected significantly upwards if the estimates of potential take due account of SOI (Silicon On Insulator) and CUBIC (CUmulatively Bonded Integrated Circuits) techniques [14] that allow three-dimensional ICs approximating biological systems to be produced. To realize such systems, however, major advances have to be made in the field of design and packaging systems, whose present trends (Fig. 14) in fact show a conflicting situation. We can expect this to be eased by a novel packaging module (Multi Chip Module) and also by the strong trend toward on-chip systems.

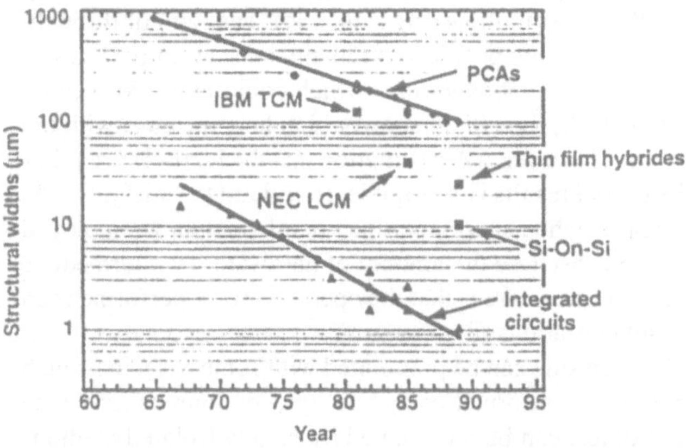

Figure 14 Conflict situation caused by "mismatch" in the development of structure width for ICs and flat modules

Finally, let us point out that, besides the forces of the "technology push" which advance Si technology primarily through leading products from standard component families, memories (DRAM) and microprocessors, other new forces described as the "market pull" will become available for driving the evolution of microelectronics forward. The origin of these new forces is to be found at the application level in the opportunities for rationalization and in novel types of system solutions. On the one hand, systems are being increasingly implemented at the chip level and thus essentially contain today's system know-how. On the other hand, tomorrow's complex hypersystems will increasingly use today's systems, and therefore future chips, as their constituent modules.

Bibliography

[1] J. Bardeen, W. H. Brattain, Phys. Rev. **74**, 230 (1948)

[2] W. Shockley, Besll Syst. Techn. J. **28**, 435 (1949)

[3] J. S. Kilby, US Patent 3 138 743 (1958)

[4] B. Noyce, US Patent 2 981 877 (30.07.1961)

[5] G. Friedrichs and Schaff, A. *Auf Gedeih und Verderb – Mikroelektronik und Gesellschaft* (1982)

[6] F. Hofmann, Hänsch, W., Geis, H., Rösner, W., Takacs, D., and Risch, L., ESSDERC '91, Montreux (1991)

[7] A. W. Wieder, GME-Fachtagung, Baden-Baden (1991)

[8] R. Burmester, Winnerl, J., and Neppl, F., Microc. Eng. 473 (1990)

[9] G. A. Sai-Halasz *et al.*, IEEE El. Dev. Lett. EDL-8, 463 (1987)

[10] H. Schwärtzel, GME-Fachtagung, Baden-Baden (1991)

[11] I. Ruge, GME-Fachtagung, Baden-Baden (1991)

[12] A. W. Wieder, ITG-Fachtagung, Stutttgart (1992)

[13] t. G. Noll and De Man, E., ISACS '92, San Diego, 1652 (1992)

[14] Y. Hayashi *et al.*Symposium on VLSI Technology, 95 (1990)

Contents of Volumes 1-33